The Handbook of Carbon Accounting

What is a "CO_2 neutral" book?

The carbon emissions resulting from the production of
this book have been calculated, reduced and offset to
render the book "carbon neutral".

The emissions related to the production of this book have been
estimated through a detailed analysis of the carbon emissions related
to the supply chain. Using research and emission factors compiled
by the French agency for the environment and energy management
(ADEME), the UK Carbon Trust, and ecoinvent V3 LCI database,
CO2logic has calculated the carbon footprint of this book.

The production of 290 g of paper is responsible for the emissions of
585 g of CO_2-equivalent (forest product manufacturing facilities, the
collection and production of the fibres, the sorting and processing of
recovered paper before it enters the recycling process). The other
processes involved in the production of this book (ink production,
transport, printing and the distribution of the book) have an estimate
carbon footprint of 325 g of CO_2 per book. In total, the carbon
footprint is estimated to be around 0.9 kg CO_2 or 2 lb CO_2 per book.
This is equivalent to driving 5 km (3 miles) with the average European

car or to working 24 hours on a desktop using the average electricity emission factor in the UK (22 hours in the US).

To improve on this result, Greenleaf Publishing uses sustainable FSC paper. Sustainably managed forests act as carbon sinks and can over time have a net positive effect on climate change. Additionally, Greenleaf is currently working to minimize and mitigate its carbon footprint, reducing waste, promoting sourcing of renewable raw materials such as wood fibre and energy, and working with its stakeholders and suppliers towards a closed-loop material and energy cycle.

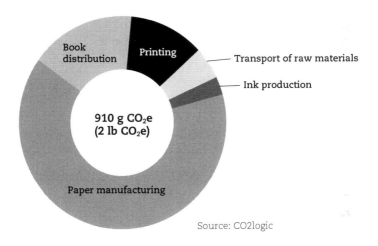

Source: CO2logic

Having calculated and analysed the options to reduce its carbon footprint, Greenleaf Publishing has concluded a partnership with CO2logic to offset the remaining emissions related to the production of this book. In practice, a project that provides efficient cookstoves in Uganda to avoid deforestation will be supported and the related carbon credits (VERs Gold Standard) will be cancelled in order to offset the relevant emissions. Through this voluntary and credible action, Greenleaf Publishing and CO2logic hope to contribute towards the protection of our climate.

THE HANDBOOK OF
CARBON
ACCOUNTING

Arnaud Brohé

Routledge
Taylor & Francis Group

LONDON AND NEW YORK

First published 2016 by Greenleaf Publishing Limited

Published 2017 by Routledge
2 Park Square, Milton Park, Abingdon, Oxon OX14 4RN
711 Third Avenue, New York, NY 10017, USA

Routledge is an imprint of the Taylor & Francis Group, an informa business

British Library Cataloguing in Publication Data:
A catalogue record for this book is available from the British Library.

ISBN-13: 978-1-78353-316-9 [hbk]
ISBN-13: 978-1-78353-317-6 [pbk]

Contents

Figures and tables

Figures

Tables

Boxes

List of acronyms and abbreviations

AAU	Assigned Amount Unit
ABC	Association Bilan Carbone (Carbon Balance Association)
ADEME	Agence de l'environnement et de la maîtrise de l'énergie (Agency for the Environment and Energy Management)
AFNOR	Association française de normalisation (French Standardization Association)
AFOLU	Agriculture, forestry and land use
ANSI	American National Standards Institute
AR4	Assessment Report 4 (from the IPCC)
AR5	Assessment Report 5 (from the IPCC)
AWG	Ad hoc Working Group
BSI	British Standards Institution
CAR	Climate Action Reserve
CAS	Centre d'Analyse Stratégique (Strategic Analysis Centre)
CBI	Confederation of British Industry

CDM	Clean Development Mechanism
CDP	Carbon Disclosure Project
CER	Certified Emission Reduction
CFC	Chlorofluorocarbon
CGDD	Commissariat général au développement Durable (General Commission for Sustainable Development)
CH_4	Methane
CO_2	Carbon dioxide
COP	Conference of the Parties of the UNFCCC
CPP	Clean Power Plan
CRC	Carbon Reduction Commitment
DECC	Department of Energy and Climate Change
DEFRA	Department for Environment, Food and Rural Affairs
DG	Directorate General
DIN	Deutsches Institut für Normung (German Institute for Standardization)
DOE	Designated Operational Entity
EC	European Commission
EPA	Environmental Protection Agency
EPD	Environmental Product Declaration
ERT	Expert Review Team
ERU	Emission Reduction Unit
EU	European Union
EU ETS	European Union Emission Trading System
EUA	European Union Allowance
EUTL	European Union Transaction Log
FAR	First Assessment Report
FEVE	European Container Glass Federation
GDP	Gross Domestic Product
GHG	Greenhouse Gas

GHGRP	Greenhouse Gas Reporting Program
GtC	Gigatonne of carbon
$GtCO_2$	Gigatonne of carbon dioxide
GWP	Global Warming Potential
HFC	Hydrofluorocarbon
IEA	International Energy Agency
IETA	International Emissions Trading Association
IFC	Institut de Formation Carbone (Carbon Training Institute)
IPCC	Intergovernmental Panel on Climate Change
ISO	International Organization for Standardization
ITL	International Transaction Log
JI	Joint Implementation
JISC	Japanese Industrial Standards Committee
JISC	JI Supervisory Committee
J-VETS	Japanese Voluntary Emissions Trading Scheme
KP	Kyoto Protocol
K-VAP	Keidanren Voluntary Action Plan
LCA	Long-term Cooperative Action
LCA	Life-cycle Assessment or Life-cycle Analysis
LCI	Life-cycle Inventory
LULUCF	Land Use, Land Use Change and Forestry
MAC	Marginal Abatement Cost
MW	Megawatt
N_2O	Nitrous oxide
NGO	Non-Governmental Organization
NO_x	Nitrogen oxides
O_3	Ozone
OECD	Organization for Economic Cooperation and Development
ODA	Official Development Aid
PDD	Project Design Document

PFC	Perfluorocarbon
PIN	Project Idea Note
PoA	Programme of Activities
Ppm	Parts per million
REDD	Reducing Emissions from Deforestation and Forest Degradation
RGGI	Regional Greenhouse Gas Initiative
SAR	Second Assessment Report (from the IPCC)
SCC	Social Cost of Carbon
SF_6	Sulphur hexafluoride
SME	Small and Medium Enterprises
SOeS	Service de l'observation et des statistiques (Observation and Statistics Directorate)
UN	United Nations
UNCTAD	United Nations Conference on Trade and Development
UNEP	United Nations Environment Programme
UNFCCC	United Nations Framework Convention on Climate Change
VCS	Verified Carbon Standard
WBCSD	World Business Council for Sustainable Development
WMO	World Meteorological Organization
WRI	World Resources Institute

Introduction

Accounting of greenhouse gas (GHG) emissions, which will be referred to as "carbon accounting" in the rest of this book, is a relatively new development. Looking back to 1988, when the Intergovernmental Panel on Climate Change (IPCC) was founded, there was an increased collective awareness of the influence of carbon dioxide (CO_2) emissions on climate. However, only in 1992, with the adoption of the United Nations Framework Convention on Climate Change (UNFCCC) during the Rio de Janeiro Earth Summit, did states begin the task of inventorying their GHG emissions. Carbon accounting among private enterprises started in the late 1990s, thanks in particular to the development of the *Greenhouse Gas Protocol*. It was not until the mid-2000s that a significant number of organisations started to calculate and publish their emission levels.

These carbon accounting initiatives were launched at different levels, from the UN to local authorities and private enterprises, but they were still not coordinated. The aims pursued differ considerably depending on the method used, but they can be grouped into two broad categories. The first of these, the "territory-based approach", also referred to as the "production-based approach", includes obligatory steps that are rigorous and often covers only direct emissions from a territory, with the aim of monitoring emissions. The second approach, the "carbon footprint", also referred to as "consumption-based approach" includes voluntary steps that are more flexible but are often less precise than those of the territory-based approach. The carbon-footprint approach has broad coverage and its basic aim is to raise awareness and encourage climate action.

As calculation methodologies have developed, scientists and economists have carried out work to estimate the scale of the damage, costs and benefits of mitigation or adaptation strategies, thereby improving our understanding of the external cost of carbon emissions. Despite their high level of uncertainty, these studies to estimate the social cost per unit of emitted carbon have helped promote application of the "polluter pays" principle, which combats climate change by imposing a carbon price on the main emitters. Once emission levels became known, thanks to the inventory methods, carbon markets and taxes developed, making it possible to integrate the value of carbon into the financial accounts of an increasing number of actors.

We will initially attempt to establish the context that led to the creation of carbon accounting. The phenomenon of the greenhouse effect, the understanding of the role of human activities and recognition of this major challenge by political leaders will be covered in Chapter 1.

We will then look at the basic principles of physico-chemical inventories of GHG emissions (Chapter 2). What are greenhouse gases? How can their totals be found and calculated?

Chapters 3 and 4 will present the two broad calculation approaches:

- the territory-based approach, which analyses direct GHG emissions, and is principally useful for producing national inventory reports

- the carbon-footprint approach, which makes it possible to calculate the carbon impact of an organisation, a territory or even a product over its lifecycle.

Chapter 5 will deal with the monetisation of CO_2, presenting the two main methods of determining the value of one tonne of CO_2: cost–benefit analysis and cost-effectiveness analysis. Two key government reports will be used to illustrate these approaches: the Stern Report (2006) on the economics of climate change, published in the United Kingdom, and the report produced by the Quinet Commission (2008) on the shadow price of carbon, published in France.

Chapter 6 will link "chemical" emissions with the financial accounts of states or enterprises, analysing the tools for integrating GHG emissions into the economic sphere. This chapter will introduce carbon taxes, the Kyoto Protocol flexibility mechanisms, the EU Emissions Trading System (EU ETS) and other similar schemes developed in North America and the Asia Pacific region. It will also cover voluntary offsetting for GHG emissions and show that implementation of emissions inventories has sometimes been seen as the first step towards creating carbon markets. Although the aim of these markets is to reduce GHG emissions, we will see that attempts by the private sector to maximize short-term profitability, and in some cases excessively lax regulation, have led to fraud and abuse of the system.

This work will demonstrate that these methods or instruments are not just a simple technique to achieve the aim of reducing emissions for the lowest cost. On the contrary, these tools generally reflect a compromise between divergent interests and individual world views. Thanks to the publication of the equivalency factors between different GHGs in 1995 and the commitment by the Parties to the UNFCCC to carry out inventories, it has been possible to create an international carbon market. This market allows a price to be put on pollution; however, many questions arise about its practical application. We will also analyse the limits of the voluntary approaches. Although carbon accounting is sometimes used by enterprises as an indicator of their dependence on fossil fuels or to show how they are affected by stricter climate regulations, many private actors use these approaches for public relations purposes or lobbying. The role played by certain companies and pressure groups in setting up carbon accounting is a further indicator that many large enterprises know that regulations are inevitable and are thus taking a proactive approach to enable them to shape legislation in their favour.

1

The birth of carbon accounting

Climate risk, the first signs of which can already be felt, truly hangs like a sword of Damocles over human development. Concentrations of GHGs have increased considerably since the Industrial Revolution, thanks largely to our methods of organization, production and consumption. In response to this challenge, the international community came together at the UN to build up a legal framework ultimately intended to "prevent dangerous anthropogenic interference with the climate system" (UNFCCC, 1992).

The greenhouse effect: a natural phenomenon amplified by human activity

The greenhouse effect is a phenomenon that maintains the earth at an average temperature of around 15 degrees Celsius°, thus allowing life to exist. It is caused by the natural presence of greenhouse gases, which trap some of the heat emitted by sun in the atmosphere. Figure 1.1 provides a brief description of the natural phenomenon.

FIGURE 1.1 **The greenhouse effect**

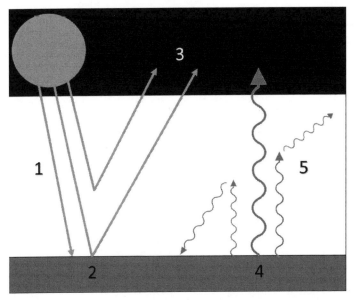

Legend: The sun supplies energy to the earth (1). This energy is partially absorbed by the earth's surface (2), but some of it is reflected back into space (3). The energy absorbed by the surface is re-emitted into the atmosphere in the form of infrared rays (IR) (4). Some of these rays are absorbed by particular gases - GHGs. These gases are quite scarce, the most commonly occurring in nature being water vapour, carbon dioxide and ozone (5). One could compare the role played by these gases to that of a greenhouse: the sun's rays penetrate the atmosphere and the GHGs ensure that the energy stays inside.

The scientific term to describe the influence of GHGs on the earth's heat balance is "radiative forcing". This is defined as the difference between the radiative energy received and the radiative energy emitted by a given climatic system. Positive radiative forcing tends to heat up the system (more energy is received than emitted), while a negative radiative forcing leads to cooling (more energy is lost than received).

The main GHG is water vapour. However, direct emissions of water vapour have virtually no impact if we look only at the anthropogenic or additional greenhouse effect (which is in addition to the natural greenhouse effect). Indeed, this planet is essentially covered in water; furthermore, water remains in the atmosphere for only around one week and thus does not accumulate. This means that water-vapour emissions of human origin are not sufficient to interfere with the global water cycle.

CO_2 is the main cause of the anthropogenic greenhouse effect. This colourless, odourless, chemical compound accumulates in the atmosphere for 125 years on average, which means that measures taken to limit emissions will not have an immediate effect on the CO_2 concentration in the atmosphere. Methane (CH_4) and nitrous oxide (N_2O) are the two other main GHGs.

Changes in human activities, which contribute to the gas emissions mentioned above, have significantly impacted the concentration of GHGs in the atmosphere. This change was first identified some time ago. As early as 1896, the chemist Svante Arrhenius observed that the level of CO_2 in the atmosphere had increased considerably since the start of the Industrial Revolution (Arrhenius, 1896). Understanding that this rise would continue with increased consumption of fossil fuels, and knowing the role of CO_2 in determining global temperatures, this perceptive Swede concluded that if the concentration of this gas in the atmosphere doubled, the temperature would increase by several degrees.

The marked rise in our consumption of fossil fuels has inevitably been accompanied by constantly increasing emissions of GHGs into the atmosphere. Oil, natural gas and coal were formed by the slow decomposition of layers of vegetable residues, capturing atmospheric carbon over many millions of years. By burning these fuels we release an additional quantity of CO_2 back into the atmosphere, interfering with the natural carbon cycle. Over several decades, we have released the quantity of CO_2 that was emitted and captured by ecosystems over hundreds of millions of years.

Before humans appeared, the earth had already set up a process whereby the oceans, forests and soils "recycled" GHG emissions (principally CO_2). However, the additional quantity released by human activities is so great that it is not all recycled by ecosystems. The IPCC (see Box 1.1) estimates that in 2000, of the 26 billion tonnes of CO_2-equivalent (tCO2e) of annual emissions of human origin, around 15 GtCO$_2$e are accumulated annually in the atmosphere and are not recycled (these emissions figures do not take into account deforestation). It is estimated that in 2012, global emissions had increased by 20% compared to 2000 (around 32 GtCO$_2$e) and deforestation continued, limiting the recycling role of carbon sinks. As an increasing proportion of emissions is left unrecycled, the concentration of CO_2 in the atmosphere increases. This number has risen from 280 parts per million (ppm) in the pre-industrial era to 403 ppm in 2016, a value that exceeds the interval of natural variation over the last 650,000 years. In 1990 this figure was just 353 ppm. Taking into account the full range of gases covered by the Kyoto Protocol, concentration of CO_2-equivalent reached 478 ppm in 2014. However, if we also include aerosols, which cool the atmosphere, concentration is reduced to 441 ppm of CO_2-equivalent, according to the European Environment Agency.

Box 1.1   The IPCC – scientific expertise to help political decision-makers

The Intergovernmental Panel on Climate Change (IPCC) was established in 1988 by the World Meteorological Organisation (WMO) and the United Nations Environment Programme (UNEP) at the behest of the G7. The role of the IPCC is to "assess the scientific, technical and socio-economic information relevant for understanding the risk of human induced climate change". This means it is not a research laboratory, but rather a body that assesses and brings together work carried out in research centres around the world.

All researchers working in an area related to the study of climate, including researchers whose work examines human influence on climate, can request that their work be studied as part of the assessment procedures organized by the IPCC. Thus far all official publications from the IPCC have been approved unanimously by the countries represented in the IPCC assembly.

In 1990 the IPCC published its first assessment report (FAR), which confirmed that climate change was a threat and called on the international community to act. The General Assembly of the United Nations responded in December 1990 by formally opening negotiations on the Framework Convention on Climate Change. The most recent assessment report (AR5) was produced in 2014 (IPCC, 2014). Almost 30 years after it was founded, the IPCC is still the most comprehensive source of information on climate change. In December 2007 its work received a Nobel Peace Prize, awarded jointly to former American vice president Al Gore.

Source: www.ipcc.ch.

In its fifth assessment report, the IPCC estimated that it was "extremely likely" (over 95% probability) that humans were responsible for the warming observed in the 20th century. In addition, the IPCC found that it was "extremely likely" that continued anthropogenic emissions over the 21st century would lead to additional warming more significant than the warming that occurred in the 20th century.

An influential paper published in Nature (Meinshausen *et al.*, 2009) shows that limiting the cumulative emissions of GHGs over the period 2000 – 2050 to 1,000 $GtCO_2e$ gives a 25% probability that warming will be over 2°C compared to the pre-industrial level (1.4°C compared to 1990). Given the total volume of emissions since 2000, we can consume only half of the economically recoverable reserves of oil, gas and coal if we are to meet this objective and limit the risk of warming being greater than 2°C. An OECD study estimates that if energy policy remains unchanged, the concentration of CO_2-equivalent in the atmosphere could reach 685 ppm before the end of the century (OECD, 2012). This study inspired the Carbon Tracker Initiative, a non-profit organization that informs investors and market regulators about the risk of stranded assets (see glossary) resulting from the study of carbon budgets. Using all fossil fuels will breach the global carbon dioxide budget, and, in this new context, capital spent on finding and developing more reserves is largely wasted (Carbon Tracker, 2013).

Sustained growth of GHG emissions

Emissions have doubled since 1970

Global GHG emissions practically doubled between 1970 and 2010. The OECD estimates that these emissions could double again over the next 40 years if no measures are taken to reduce emissions, largely as a result of the particularly significant growth in the electricity sector (OECD, 2012).

FIGURE 1.2   Changes in global greenhouse gas emissions (CO_2-equivalent) emissions since 1970

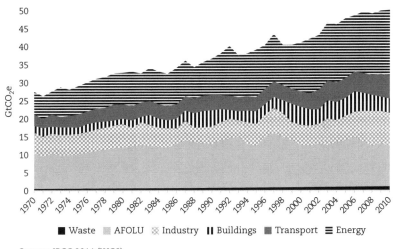

■ Waste ▦ AFOLU ▨ Industry ‖ Buildings ■ Transport ≣ Energy

Source: IPCC 2014 (WG3)

China has been the largest emitter in the world since 2007, following spectacular growth in emissions. Emissions by the United States, the second largest emitter, have recently fallen, principally due to a transition from coal to gas in the electricity-generation sector. In the European Union (EU), emissions by the first 15 member states stabilized between 1990 and 2005, with a slight increase in CO2 emissions that was cancelled out by a reduction in emissions of CH4 and N2O. The reduction in emissions has been more marked since 2009, principally due to the slowdown in industrial production. The reduction is even more marked for the 27 member states, due to the brutal transition undergone by the countries of the former eastern bloc since the start of the 1990s. As a result of the increase in emissions by China between 2000 and 2011, emissions in China (7.2 tCO2e) and in the EU27 (7.5 tCO2e) were comparable by 2011 (see Figure 1.3).

FIGURE 1.3   Changes in global emissions in the three main global economies (in tCO₂/capita)

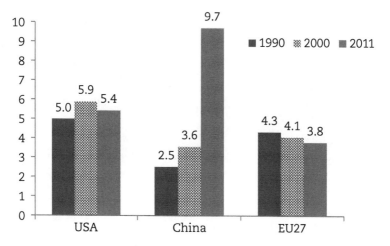

Source: European Commission, Joint Research Centre (JRC)/PBL Netherlands Environmental Assessment Agency. Emission Database for Global Atmospheric Research (EDGAR), release version 4.2

Various sources

Global anthropogenic emissions of GHGs can be divided into six broad categories (see Figure 1.2):

1. **Energy generation** (electricity and heat) is responsible for one-third of emissions. Absolute emissions almost trebled to reach 17 GtCO₂eq/year in 2010.

2. **Industry** is responsible for nearly one-fifth of global GHG emissions. Its direct emissions (excluding waste, waste water and AFOLU contributions) grew from 5.4 $GtCO_2eq$/year in 1970 to 8.8 $GtCO_2eq$/year in 2010. The industrial-processes sector includes GHG emissions produced by various industrial activities not linked to energy, in which materials are processed chemically or physically. In industrial processes, such as cement production, metal and steel smelting in blast furnaces and the production of adipic acid, various GHGs can be emitted, such as CO_2, CH_4 and N_2O.

3. **Building** emissions grew from 2.5 $GtCO_2eq$ in 1970 to 3.2 $GtCO_2eq$ in 2010 with emissions-growth rates in OECD-1990 countries being largely negative.

4. **Transport** emissions grew from 2.8 $GtCO_2eq$ in 1970 to 7 $GtCO_2eq$ in 2010.

5. Emissions from **agriculture, forestry and other land-use** (AFOLU) are associated with livestock farming, manure and agricultural land as well as land-use changes. Emissions from livestock are caused by enteric fermentation (ruminants produce significant quantities of methane during their microbial digestion process). Land-use emissions covers emissions from man-made fires and controlled burning, as well as conversion of forests or prairies into agricultural land. AFOLU related emissions increased by 20% from 9.9 $GtCO_2eq$ in 1970 to 12 $GtCO_2eq$ in 2010, constituting about 20% to – 25% of all global emissions in 2010. This sector can be divided in two subsectors of approximately equal size: agriculture and land-use change

(mostly deforestation, which is continuing at a sustained rate in developing countries).

6. The **waste** sector includes emissions produced by treatment and disposal of waste. Sources of emissions include disposal of waste in the soil (dumps), waste-water treatment and incineration of waste. This sector emits barely 2.5% of total anthropogenic greenhouse gas emissions in 2010.

With the exception of deforestation, these broad categories apply to high-income regions such as Europe, Japan and North America. Deforestation does not pose a problem in these regions; there has been slight reforestation since 1990.

Low and lower middle-income countries contribute the largest share of emissions associated with AFOLU. Between 2000 and 2010, emissions by upper-middle-income countries from energy ($+3.5$ GtCO$_2$e/year) and industry ($+2.4$ GtCO$_2$e/year) more than doubled, and by 2010, emissions from industry in upper-middle-income countries have surpassed those from high-income countries (IPCC, 2014).

Transport and building account for a more significant proportion of emissions in Europe and North America (over one-third of GHG emissions) than globally (one-fifth).

While AFOLU emissions represent a small percentage in developed countries, almost 90% of GHG emissions from 1970 – 2010 in the least developed countries (LDCs) were generated by this sector alone.

Climate change affects many ecosystems

The climate change that is currently underway, particularly temperature rises, have already influenced many physical systems and it has an impact on biodiversity and the way our societies are organized.

The IPCC estimates that sea levels will rise between 26 and 82 cm during the period from 1986 – 2005 to 2081 – 2100 (IPCC, 2014). This rise in sea levels will threaten coastal areas and will necessitate significant adaptation initiatives. The rise in water levels will be especially problematic for states with territories slightly above sea level, particularly small island states and Bangladesh, where millions of people could be displaced. In addition, these territories lack sufficient capital to finance the infrastructure needed to combat these changes. It is also highly likely that certain extreme events will become more frequent and/or more intense (particularly heavy rains, cyclones, heatwaves and droughts).

The scientific community has stressed how vulnerable many unique systems are to these changes (for example glaciers, coral reefs and atolls, mangroves, boreal and tropical forests, polar and alpine ecosystems, humid areas of the prairie and remaining native grasslands). Scientists also predict that climate change will increase the risk of extinction for certain species.

This combination of phenomena will have a significant social and economic impact, particularly through consequences related to food production, access to water and health. Many human societies will be forced to adapt by modifying their organizational structures, and the increase in the number of climate refugees is expected to be significant but difficult to quantify (Biermann and Boas, 2010). The mortality rate will also rise in the most vulnerable regions due to an increase in illnesses and lack of access to water. In addition, the highly likely increase in the number of droughts and other extreme climatic phenomena will have an impact on the risk and scale of humanitarian crises. Hundreds of millions of people could suffer from hunger, lack of water and coastal flooding as the planet heats up (Stern, 2006).

Political recognition of the problem: from Rio to Doha

The UNFCCC aims to stabilize concentrations at a level that would prevent dangerous interference

On 9 May 1992 the United Nations Framework Convention on Climate Change (UNFCCC, 1992) was adopted. It was presented for signatures during the UN conference on the environment and development, the Earth Summit, in Rio de Janeiro, on 4 June 1992, and it came into force on 21 March 1994, after being ratified by 50 states. The ultimate objective of the Convention (article 2) is the "stabilization of greenhouse gas concentrations in the atmosphere at a level that would prevent dangerous anthropogenic interference with the climate system". This level, which is not defined by the Convention, should allow ecosystems to adapt naturally to climate change, ensure that food production could be maintained and enable sustainable economic development. Although these non-quantitative commitments from international agreements have inspired widespread support, it is unlikely that they will create the necessary conditions for states to change their behaviour.

The Convention divided countries into two groups: those listed in Annex I (Annex I Parties) and those not listed there (non-Annex I Parties). Annex I Parties are the industrialized countries who have made the largest historical contribution to GHG emissions. Their emissions per inhabitant are higher than those of most developing countries and they have the advantage of the financial and institutional means to face the problem. The principles of equity and "common but different responsibilities" guaranteed in the Convention demand that these nations set an example by changing their long-term emission trends. To this end, the Annex I Parties made a commitment to adopt measures and national policies with the (legally non-binding) aim of reducing their emissions to 1990 levels by 2000. Non-Annex I countries are generally developing countries. However, various recently industrialized countries, such as South Korea, China, Mexico and South Africa, fall under this category. It is important to stress that the Convention commits all Parties (Annex I Parties and non-Annex I Parties) to produce inventories of their GHG emissions.

In 2016, 196 states and the EU were Parties to the Convention. The Parties meet annually during the Conference of the Parties (COP), the supreme decision-making body of the Convention. During these meetings the Parties make the necessary decisions to promote effective application of the Convention.

The Kyoto Protocol sets binding objectives

During the first Conference of the Parties (COP 1), which took place in Berlin in 1995, the Parties found that the specific commitments in the Convention for the Annex I countries were not sufficient as they were too vague (see Box 1.2, "The Conferences of the Parties to the UNFCCC", which presents a summary of the main advances at each COP). The Parties then launched a new cycle of discussions to reach stricter and more precise commitments for the Annex I countries. After two-and-a-half years of intense negotiations, the Kyoto Protocol was adopted during the third COP on 11 December 1997 in Japan and entered into force on 16 February 2005.

This protocol breaks new ground in international law in several respects. First, inspired by the success of the Montreal Protocol (which relates to the protection of the ozone layer), this text sets binding and measurable objectives for the countries that ratify the text, setting it apart from the statements of intent that often characterized international law on the environment. This protocol is also the first international application of a cap and trade emission-rights system. The Kyoto Protocol enables for a GHG emissions market that distinguishes between initial allocations (which are supposed to be fair) and final allocations (modified by economic exchanges). The initial period of commitment, which affects only the countries listed in Annex B of the Kyoto Protocol (see glossary), limits the level of emissions for the period 2008 – 2012 in relation to the reference year 1990 (Kyoto Protocol, 1997).

On 3 December 2007, the new prime minister of Australia, Kevin Rudd, signed the papers for Australia's ratification of the Kyoto Protocol, making the United States the only developed country that was a signatory to the Protocol but had yet to ratify it (both Republicans and Democrats rejected the measure in a parliamentary vote). In January 2012, 192 states and the EU were Parties to the Convention. However, at the end of 2011 Canada withdrew from the Protocol, announcing unilaterally that it would no longer respect its commitments.

The flexibility mechanisms of the Kyoto Protocol, which led to the creation of an international carbon market, will be presented in detail in Chapter IV.

The difficulty of regaining the initiative

The Conferences of the Parties following the negotiation of the Kyoto Protocol were essentially dedicated to the interpretation and concrete implementation of the Convention and its Protocol. COP 13 in Bali marked the tenth anniversary of Kyoto, and it was the first conference in which negotiations focused on providing a follow-up to the Kyoto objectives, as they were set to expire at the end of 2012. Bali ended by setting a timetable to achieve a follow-up agreement to the Protocol at the latest by December 2009, during COP 15 in Copenhagen. This timetable proved to be too ambitious because of disagreements between the Parties, notably between the two main GHG emitters, China and the United States. The agreement finally reached in Denmark was minimalist and did not set precise objectives. However, Copenhagen did make tentative progress on two points. On one hand, it was the first time that China and the United States signed an agreement intended to limit temperature rises to 2°C (even though this was merely a pious hope without binding objectives on emissions levels). To reach this target, scientists estimate that global GHG emissions need to be reduced by 40 – 70% by 2050 and that carbon neutrality (zero emissions) needs to be reached by the end of the century at the latest (Meinshausen *et al.*, 2009).

The agreement also promised significant aid to allow the poorest countries and those most affected by global warming to adapt. This programme, known as Fast Start Finance, is intended to mobilize an additional $30 billion for mitigation or adaptation policies in developing countries. Developing countries have accepted even more ambitious medium-term financing objectives, with $100 billion per year to be mobilized by 2020 (Bodansky, 2010). In the context of an economic crisis and a debt crisis in most "donor" countries, the question of additionality, particularly in terms of public development aid, and the conditions of this financing are key concerns for developing countries.

The Cancun Agreements of December 2010 made it possible to integrate the results of the Copenhagen Summit of 2009 into the UN process, particularly the idea of creating a Green Climate Fund to channel a significant proportion of the additional financing promised in Copenhagen. At Cancun it was also possible to adopt the principle of creating new institutions to promote technology transfer and coordination of adaptation policies. Finally, there was also progress on the extension of the economic tools from the Kyoto Protocol, a carbon finance mechanism linked to deforestation and reforestation (REDD+), and the creation of new economic tools, although no formal decisions were made.

COP 17, which was hosted by Durban at the end of 2011, was not held under the most auspicious circumstances. Indeed, the year was signified economically by the continuing financial crisis, and politically by announcements that Canada, Russia and Japan would not participate in a second commitment period. The absence of significant emitter countries such as the United States and China also loomed. The EU, for its part, made its participation conditional on the definition of a roadmap for a new climate architecture by 2015. The Durban conference managed to get an agreement back on track for 2015, which should have been obtained in Copenhagen in 2009.

The three following COPs (in Doha, Warsaw and Lima) closed without major breakthroughs but laid the groundwork for the decisions that were to be made in Paris.

The Paris Agreement: a new momentum for climate action?

COP 21 in Paris concluded with an agreement to accelerate climate action. This agreement represents a significant paradigm shift. The focus has now moved away from the top-down policy architecture and market-based approach that was initiated by the Kyoto Protocol in favour of a bottom-up architecture where governments set targets at a national level and adopt various policy approaches. Not all of these policy approaches are market-based. Indeed, the Paris Agreement requires all Parties to put forward their best efforts through "nationally determined contributions" (NDCs) defined at a country level. There will also be a global stocktake every five years, which allows the nations to analyse their collective climate action by assessing the progress that has been made toward achieving the purpose of the agreement.

The Paris Agreement's central aim is to strengthen the global response to the threat of climate change by taking the actions necessary to limit global temperature rise this century to below 2°C above pre-industrial levels and to pursue efforts to limit the temperature increase even further to 1.5°C (Paris Agreement, 2015). Additionally, the agreement aims to reinforce the ability of countries to adapt to the impacts of climate change. To reach these ambitious goals, appropriate financial flows, a new technology framework and an enhanced capacity-building framework will be put in place. These measures will support developing countries and the most vulnerable countries as they implement climate actions. The agreement also enhances transparency.

The Paris Agreement represents an effort to move beyond the mandatary treaty-based approach of the Kyoto Protocol and the purely voluntary approach of the Copenhagen Accord, but it is still uncertain what framework of agreement the participating nations will choose. On the one hand, the Paris Agreement went to great lengths to avoid binding goals, and compliance with pledged carbon-emission reduction is a voluntary undertaking. On the other hand, countries are signing the accord under the assumption that the agreement will not enter into force until enough states have both signed and ratified the agreement. This suggests that the Parties consider the Paris Agreement to be a treaty.

The Paris Agreement opened for signature on 22 April 2016 – Earth Day – at the UN headquarters in New York. The agreement will enter into force 30 days after 55 countries that account for at least 55% of global emissions have deposited their instruments of ratification. This process can take several years (it took seven years for the Kyoto Protocol). As of November 2016, 193 UNFCCC members have signed the treaty, 105 of which have ratified it. After the EU ratified the agreement in October 2016, there were enough countries that had ratified the agreement that produce enough of the world's greenhouse gases for the agreement to enter into force. The agreement went into effect on 4 November 2016. The recent election of climate-sceptic Donald Trump as President of the USA has jeopardized the USA's commitment to the agreement and called its overall effectiveness into question.

Despite the progress of having a global agreement, the bottom-up approach and current plans (NDCs) are still far from enough to keep global warming to tolerable levels. To achieve these levels it is necessary to transition from a fossil fuel–based energy system to an entirely clean energy system no later than the middle of this century. Substantial enhancement or over-delivery on current NDCs by additional national, sub-national and non-state action is required to maintain a reasonable chance of meeting the Paris target (Rogelj, et al., 2016). Unless countries develop more ambitious plans, the world could ultimately suffer the profound consequences described above.

Box 1.2   The Conferences of the Parties of the UNFCCC

1995 – COP 1: Berlin. The first Conference of the Parties after the Convention came into force.

1996 – COP 2: Geneva. The ministerial declaration, which was noted (but not adopted), called for definition of "legally binding" intermediate objectives.

1997 – COP 3: Kyoto. Adoption of the Convention Protocol defining the legally binding objectives for Annex I.

1998 – COP 4: Buenos Aires. Little progress on implementation of the Protocol.

1999 – COP 5: Bonn. Technical meeting. Founding of the Consultative Group of Experts on "National Communications" for non-Annex I Parties.

2000 – COP 6: The Hague. Various disagreements among the Parties, particularly regarding the United States' proposal to grant credits for carbon sinks and financing for adaptation measures in developing countries. The meeting ended in failure, and it was suspended without agreement.

2001 – COP 6 Part 2: Bonn. An intermediate meeting held in July 2001 in Bonn, intended to revive dialogue.

2001 – COP 7: Marrakesh. Conclusion of the "Marrakesh Accords". The principal rules for international exchange of emissions rights and practical aspects of the other flexibility mechanisms were set at this COP.

2002 – COP 8: New Delhi. Discussions mainly dealt with Russia's reluctance to sign the Kyoto Protocol.

2003 – COP 9: Milan. Clarification of the use of the Adaptation Fund established in 2001 in Marrakesh.

2004 – COP 10: Buenos Aires. Initial reflections on the post-2012 framework.

2005 – COP 11: Montreal. The first meeting of the Parties (MOP 1) following ratification of the Kyoto Protocol. The Montreal Action Plan envisaged that Kyoto would still be followed after 2012 and defined more ambitious objectives for a second period of commitment. The Ad Hoc Working Group on the Kyoto Protocol (AWG KP) was set up in Montreal.

2006 – COP 12: Nairobi. Progress was principally made on the Adaptation Fund and the functioning of the Clean Development Mechanism.

2007 – COP 13: Bali. The Ad Hoc Working Group on Long-term Cooperative Action (AWG-LCA) was set up in Bali. Its work plan, known as the "Bali Road Map", covered mitigation, adaptation, technology transfer and financing.

2008 – COP 14: Poznan. Discussions focused on post-2012, and the delegates left hoping to reach an agreement during the next COP.

2009 – COP 15: Copenhagen. Despite the presence of more than 60 leaders, including the presidents of the United States, China and India and the majority of the European leaders, the Copenhagen Summit did not achieve an ambitious binding agreement. It was in Copenhagen that Parties stated their determination to limit global warming to 2°C between now and 2100.

2010 – COP 16: Cancun. The Cancun Agreements of 2010 made it possible to integrate the results of the Copenhagen Summit of 2009 into the UN process, notably the idea of creating a Green Climate Fund. Japan announced that it will not sign any extension of the Kyoto Protocol.

2011 – COP 17: Durban. Clarification of the use of the Adaptation Fund established in 2001 in Marrakesh.

2012 – COP 18: Doha. Still no clarification of the climate financing promised in Copenhagen. The delegates agreed to extend the Kyoto Protocol (but with fewer Parties as some key countries pulled out, notably Canada and Japan).

2013 – COP 19: Warsaw. Further progress was made as the Warsaw Mechanism was proposed. This would provide aid, expertise and funding to developing countries affected by extreme weather events.

2014 – COP 20: Lima. The discussions centred around the green economy paradigm, and "the Lima call for climate action" laid the groundwork for the decisions that were to be made in Paris.

2015 – COP 21: Paris. By consensus, the 195 participating countries agreed to the final form of the Paris Agreement. The ultimate goal of the agreement is to reduce global greenhouse gas emissions enough to bring the global mean temperature increase below 2°C compared to pre-industrial levels.

2

The basic principles of carbon accounting

Colourless and odourless, CO_2, the gas that bears most responsibility for the anthropogenic greenhouse effect, is completely imperceptible to our senses. The same applies for most other GHGs. For this reason it is only through reliable inventories at regular intervals that we can quantify the scale of the source of the problem and determine the actions necessary to limit anthropogenic influence on the composition of the atmosphere and on climate.

A GHG emissions inventory is a document that lists and quantifies the emissions for an entity (a state, a local authority, an enterprise or organization) during a given period (generally a year) or for a product, based on an analysis of its climatic impact over its lifecycle (from design to end-of-life waste processing). This type of inventory makes it possible to identify emissions sources and to calculate which are most significant, so that the entity carrying out the study can prioritize them. Such inventories also give third parties (other states, clients, NGOs etc.) a better understanding of the context for reduction strategies and enable them in turn to evaluate the results of the policies and measures implemented by the entities publishing an inventory.

The first sections of this chapter will present the unit of measurement for carbon accounting, the tonne of CO_2-equivalent. It will then examine two particular cases: accounting of use, storage or destruction of biomass, and second, radiative forcing in the aviation sector. Finally, the basic rules and principles of carbon accounting will be introduced.

Units of measurement

Tonnes of CO_2-equivalent as a yardstick

Anthropogenic GHG emissions are mainly carbon dioxide (CO_2), methane (CH_4), nitrous oxide (N_2O), refrigerant gases (HFCs, PFCs and CFCs), sulphur hexafluoride (SF_6), water vapour (H_2O) and ozone (O_3). A kilogram of GHGs released into the atmosphere has a different radiative effect depending on the type of gas. Each GHG has a specific global warming potential (GWP) which quantifies its impact on climate during a certain period. This is not an absolute value, but rather relative to a given mass of CO_2. The potential warming effects of GHGs are translated into CO_2-equivalent using the GWP of each gas.

GWP_N represents the ratio between the cumulative radiative forcing, that is, the radiative power that the GHG sends toward the earth over a period of N years, and the same figure for CO_2.

The higher the GWP_N, the greater the additional greenhouse effect caused by releasing a specific mass of this gas into the atmosphere in comparison to the same mass of CO_2 over N years. Since the GWP_N always compares GHGs to CO_2, the GWP_N of CO_2 is by definition always going to be 1, whatever the value of N. However, this means that changes in the understanding of the global-warming impacts of CO_2 will result in changes to the GWP of all other GHGs as well.

Unless otherwise stated, the period N is set at 100 years by convention. The IPCC also provides GWP estimates for 20 years and 500 years. However, 100 years is closer to the average atmospheric lifetime of CO_2. Note that carbon dioxide has a variable lifetime: although more than half of the CO_2 emitted is removed from the atmosphere within a century, some fraction of emitted CO_2 remains in the atmosphere for many thousands of years (IPCC, 2007).

Multiplying masses of GHGs by their GWP gives a unit of measurement (tonnes of CO_2-equivalent, which is abbreviated to tCO_2e or t eq CO_2) that enables us to compare the contribution of these different gases to global warming.

There are significant areas of uncertainty relating to GWPs; however, the values published by the IPCC are the most authoritative. In 1990, the IPCC provided the GWP for 19 gases (the results were preliminary only). In 1995, the list was extended to include 26 gases. HFC-23, which was to become famous in the context of the Clean Development Mechanism (CDM) of the Kyoto Protocol, was included for the first time, with a $GWP_{100 \text{ years}}$ of 11,700 (see Table 2.1). As the $GWP_{100 \text{ years}}$ of methane is 28 according to the latest estimates, we calculate that over 100 years, the climatic impact of one tonne of methane is equivalent to that caused by the emission of 28 tonnes of CO_2 (1 t CH_4 = 28 tCO_2e). We should note that the time a gas remains in the atmosphere depends on the prevailing conditions: for example if the sinks that absorb carbon dioxide are saturated, it will remain in the atmosphere longer. This means that GWPs may change depending on the saturation levels of these carbon sinks.

Indeed, determining the warming potential of different GHGs is a complex science that is constantly evolving. As a result, the concept of GWPs has been the subject of criticism, particularly by sociologists investigating the consequences of using equivalency factors that are calculated and controlled by a handful of scientists (MacKenzie, 2009). Cumulative radiative forcing cannot be determined by releasing a kilogram of CO_2 and observing what happens over the course of a century. It is a mathematical function of a model simulating exchanges of carbon between the atmosphere, the oceans and the terrestrial biosphere. These models use the results of complex spectroscopic studies and modelling (particularly for temperatures).

The models and databases used to determine GWP are principally managed by the Harvard-Smithsonian Center for Astrophysics. In MacKenzie's view, GWPs are "black boxes" in the sense of Callon and Latour (1981). One critical question raised about the technique is whether GWPs as defined by the IPCC are really the best way to represent climatic effects. One might also question the choice of a 100-year period in carbon accounting as established by the United Nations and followed by all approaches (territory-based approaches and carbon-footprint approaches, both voluntary and obligatory).

Taking a period of 20 years or 500 years gives very different values. For example, according to the IPCC (2007), the $GWP_{20\ years}$ of methane is 72, and the $GWP_{500\ years}$ is 7.6. Carbon accounting that analyses the climatic impact of GHGs over a period of 20 years would give very different results compared to those provided by the conventional approach using a century (over 20 years, agriculture sources account for the majority of emissions). Another criticism is that the level of uncertainty is extremely high, around ±35%, according to the IPCC. For example, between 1995 and 2007, the $GWP_{100\ years}$ of HFC-23 increased by 26% (from 11,700 to 14,800) and the figure for methane increased by 19%

(from 21 to 25). Given the impact of this type of change on carbon markets that include different GHGs, one can imagine the pressure that might come to bear on the scientists responsible for determining these factors.

The revision of GWPs can be mostly explained by two main factors:

- Change to the inclusion and strength of indirect effects

- Changes in concentration of other GHGs

For instance, in the IPCC's Fifth Assessment Report (AR5) the changes to the GWP for methane is mostly explained by the attribution of CH_4 to the indirect radiative forcing effect of the ozone created in the troposphere. This reaction with CH_4 increased the GWP of CH_4 by 50%.

Table 2.1 shows the changes in GWPs over the last four reports published by the IPCC. We can see that in its last report (AR5), the IPCC published two sets of GWP values for the first time: one that accounts for climate-carbon feedbacks, which measure the indirect effects of changes in carbon storage due to changes in climate; and one that does not.

The question then becomes, which one should we use in carbon accounting?

In practice, the adoption of new GWP values has been fairly slow and cautious so far, as national and international agencies emphasize year-over-year consistency and coherence between reporting parties' inventories. Even in 2016, organizations conducting their own GHG assessments typically use AR4 GWPs, unless they previously used the IPCC Second Assessment Report (SAR) GWPs and wish for subsequent reports to be comparable.

The SAR coefficients that were published in 1995, before the Kyoto negotiations, were used for emissions calculations for

TABLE 2.1   Global warming potential of different GHGs according to the IPCC

Gas name (formula)	GWP$_{100 \text{ years}}$ IPCC 1995 (SAR)	GWP$_{100 \text{ years}}$ IPCC 2001 (TAR)	GWP$_{100 \text{ years}}$ IPCC 2007 (AR4)	GWP$_{100 \text{ years}}$ IPCC 2013 (AR5) without climate feedback	GWP$_{100 \text{ years}}$ IPCC 2013 (AR5) with climate feedback
Carbon dioxide (CO_2)	1	1	1	1	1
Methane (CH_4)	21	23	25	28 30 (fossil)	34
Nitrous oxide (N_2O)	310	296	298	265	298
Tetrafluoro-methane (CF_4)	6,500	5,700	7,390	6630	7350
Sulphur hexafluoride (SF_6)	23,900	22,200	22,800	23,507	26,087
HFC-23 (CHF_3)	11,700	12,000	14,800	12,397	13,856
HFC-125 (C_2HF_5)	2,800	3,400	3,500	3,169	3,691
HFC-134a (CH_2FCF_3)	1,300	1,300	1,430	1,300	1,550
HFC-143a ($C_2H_3F_3$)	3,800	4,300	4,470	4,805	5,508
HFC-152a ($C_2H_4F_2$)	140	120	124	137	167
HFC-227ea (C_3HF_7)	2,900	3,500	3,220	3,348	3,860
HFC-236fa ($C_3H_2F_6$)	6,300	9,400	9,810	8,056	8,998

national inventories and project mechanisms until 2012. At the governmental level, the values compiled in the IPCC's AR4 in 2007 have been recognized only at the 2011 Durban conference. National inventories have only been calculated with these new values since January 2015, more than seven years after their release. The US Environmental Protection Agency's GHG Rule required the application of AR4 GWPs as of January 1, 2014, in order "to better characterize the climate impacts of individual GHGs and to ensure continued consistency with other U.S. climate programs". Though some commentators urged the EPA to adopt AR5 GWPs, this was ultimately rejected due to concerns that the AR5 GWPs would not be widely applied for many years.

For organizations that wish to adopt the new GWP values in their voluntary GHG assessments, it is important to highlight which emission changes should be solely attributed to modifications in GWPs, as opposed to variations in activity levels. If a significance threshold is surpassed, it may also be necessary to conduct baseline recalculations in order to maintain comparability and compliance with reporting standards. The GHG Protocol, which provides the most widely used and respected standard, released an amendment to its "Corporate Standard" that recommends entities to "use GWP values from the most recent Assessment Report", but goes on to say that they "may choose to use other IPCC Assessment Reports". Since this amendment was released in February 2013, one year before the release of the IPCC's AR5, there is no indication of whether or not the AR5's GWP values with or without feedback effect are considered preferable. Several other authorities provide signs that point toward an increasingly widespread use of GWP values without feedback. France's "Agence de l'Environnement et de la Maîtrise de l'Énergie" (ADEME), for example, applies the GWPs without feedback in the latest update to its "Base Carbone" tool. The EPA's 2014 U.S. GHG Inventory also notes that AR5

GWP values without feedback have a calculation methodology that is more consistent with those that were used in the AR4.

Biogenic carbon and carbon sinks

Carbon accounting makes a distinction between fossil carbon and carbon from living matter. *Biogenic carbon* always refers to living carbon, found in all animals and vegetables. Biogenic carbon emitted during combustion of renewable energy sources is considered to be CO_2-neutral, and it is entered into inventories with a value of zero. As flora and fauna are carbon sinks, if the source is renewable and sink capacity is not reduced, we can regard living matter as carbon neutral and consider the emissions released at the end of life to correspond to the carbon trapped during growth of the flora or fauna.

For most forests in the EU, use of biomass is considered to be CO_2-neutral as European forests regenerate and are even slightly growing. However, we cannot consider use of biomass from most tropical forests to be neutral, as new trees are not planted to replace those that are cut down (this must be the case as total forest cover is diminishing). Use of tropical wood also leads to significant additional emissions. Indeed, in order to make money from some commercially lucrative species, loggers cut roads which are then used by peasants to clear the rest of the forest, causing significant CO_2 emissions.

Most GHG-emissions-accounting methodologies, including the GHG Protocol Product Standard and the PAS 2050, consider wood to be a "carbon sink" if it is used as a working material or for applications with a life-span of over a year. As a result, these methodologies credit such carbon stored beyond one year with negative emissions. It is true that the wood contains carbon that was extracted from the atmosphere during the growth of the tree. So long as the carbon contained in the trees that are cut down remains trapped and is not released into the atmosphere, and at the same time other trees grow to replace the ones that have been felled, forestry activities help extract CO_2 from the atmosphere. However, if new trees are not planted (or there is no natural regeneration), cutting down a tree to make a product simply moves an existing stock without building up any new stock (CO_2-neutral). Note that the aforementioned standards recommend to report negative emissions from stored carbon separately.

In theory, carbon sinks offer a way to help quickly clear the atmosphere of CO_2. However, in practice reforestation and changes in agricultural practices are difficult to implement, and the results are difficult to measure. For example, to estimate the impact of planting a forest, it is necessary to know what the forest is replacing, which type of soil is being reforested, what is done with the biomass that was previously there, which climatic zone is involved, which species of tree is planted, etc. This means increased uncertainty about whether the forests will continue to be carbon sinks for the forthcoming decades as the climate changes. For example, if freshly planted forests will soon decline due to climate change, it becomes debatable whether an emissions credit should be granted for them today. Eventually, all of the carbon finds its way back into the atmosphere when a tree dies or burns. Finally, the understanding of the impact of trees on climate is still uncertain. For instance, trees emit reactive volatile gases. Chemical reactions involving tree VOCs (volatile organic compounds) produce methane and ozone, two powerful greenhouse gases, and form particles that can affect the condensation of clouds.

The particular case of the aviation sector

In addition to the CO_2 emitted during combustion of kerosene, aircraft emit NO_x, water vapour and aerosols or soot that contribute to the emission of GHGs such as ozone and the formation of condensation trails or cirrus clouds, which influence the radiative forcing on the earth. Another factor is that the same quantity of emissions of one of these pollutants can have a significantly different effect depending on the altitude at which it is emitted, the season of the year or the time of day or night. It is clear, therefore, why the climatic impact of the sector is difficult to estimate.

To describe the overall radiative effect of air transport emissions at a given time, some agencies, for instance UK DEFRA or French ADEME, recommend applying a multiplying coefficient to the effect of the cumulative CO_2 emissions. In its 2015 GHG emission factors hub, the US EPA's Center for Corporate Climate Leadership provides emissions that are based on work from DEFRA but without considering a multiplying factor for radiative forcing.

Although such multiplying coefficients are a first step toward estimating overall impacts, exogenous factors (altitude, time and season) mean that they are still imperfect. The IPCC estimates the total impact of emissions from the aviation sector to be between two and four times higher than the warming effect due to CO_2 emissions alone with an average at 2.7 (IPCC, 1999). More recently, Sausen *et al.* (2005) reviewed this figure down to 1.9, based upon better scientific understanding, which mostly reduced the contrail radiative forcing.

Still it is worth remembering that the phenomena presented above have very different life-spans. The life-span of CO_2 in the troposphere and the stratosphere (where some intercontinental flights take place) is approximately 100 years. The duration of other phenomena is much shorter. The life-span of emissions other than CO_2 in the troposphere is two weeks or less – the life-span of ozone produced by nitrous oxides is about a month. It stays longer in the stratosphere, but in any case it is less than ten years. However, although emissions other than CO_2 have a relatively short life-span, they have powerful effects. We should note that a short-lived effect that is constantly renewed for a century will also have a significant effect on the climate. The only difference is the immediate effectiveness or ineffectiveness of measures taken against the pollutant. If CO_2 emissions were stopped today, they would continue to influence the climate for another century, but if we stopped NO_x emissions, the radiative forcing caused by ozone creation would stop immediately.

While currently there is no suitable climate metric that expresses the relationship between emissions and the radiative effects from aviation in the same way that the global warming potential does, this is an area of ongoing research. Nonetheless, it is clear that aviation imposes other effects on climate that are greater than that implied from only considering its CO_2 emissions. For the time being, multiplying aviation CO_2 emissions by a factor of 1.9 is recommended. The practical consequences are in fact minor as regards informing airlines and people on climate action; whatever the size of the factor used, the easiest way for most companies to lower emissions is to invest in more efficient aircraft and purchase less carbo-intensive fuel. People will find that the best way to reduce their carbon footprint is to cut back on flying.

Calculation or measurement? The role of emission factors

Given the significant number of sources of GHG emissions and their very low concentration in the atmosphere, it is rarely possible to measure GHG emissions resulting directly from a particular action. For a precise measurement of emissions resulting from human activities, we would need to put sensors on all our chimneys and vehicles, and also measure the diffuse sources of methane at dumps, cattle farms, paddy fields and so on. This approach is clearly not feasible.

For this reason it is necessary to calculate *estimates* of these emissions, based on *activity data*. Inventories can be made using activity data such as the number of litres of diesel consumed by a fleet of vehicles, the number of tonnes of iron ore used in an industrial process or, in the case of a farm, the number of cattle.

This activity data is then multiplied by the *emission factors* to determine the total quantity of GHGs emitted by that particular activity. An emission factor, also referred to as a conversion factor, shows the quantity of GHGs emitted in relation to a particular activity, in CO_2-equivalent per accounting unit of the activity (for example: kg CO_2-equivalent/kWh, kg CO_2-equivalent/tonne of steel, kg CO_2-equivalent/cow-year etc.).

The emission factor is thus the ratio between the quantity of GHG emitted by an activity, object or material and the characteristic value of the activity, object or material, measured in the unit that best defines it. A well-known example is the quantity of CO_2 emitted per kilometre of car travel, measured in grams of CO_2 per kilometre.

The emission factors used for the calculation include a margin of uncertainty that is sometimes far from negligible. For example, the levels of uncertainty for emission factors listed in the French database Base Carbone are estimated to be between 5% and 50%. If the levels of uncertainty for emission factors linked to combustion of hydrocarbons is low (about 5%), the opposite is the case for levels of uncertainty related to complex manufactured products (such as computers) or food products, which are often between 25% and 50%.

The emission factors used in inventory methods are often the result of approximations or strong assumptions. They reflect average values or sometimes state-of-the-art values for technical equipment. It is clear that every electric kilowatt hour, every sheet of steel and every cow has a different emission factor depending on individual circumstances. Although emission factors relate to an activity, they are often a function of current practices. By their very nature, these factors change, so when carrying out a GHG emissions inventory it is vital to choose recent emission factors that are appropriate for the relevant context.

Emission factors can also be highly mutually dependent. For example, if the proportion of renewable energy increases, the emission factor of electricity will be reduced. This reduction will have an effect on the emission factors of all the products and services that use electricity during production: the emission factor of public transport companies using electric vehicles, emission factors for semiconductors, emission factors for use of air conditioning etc. Table 2.2 provides examples of emission factors.

Primary and secondary data

Primary data is quantitative data gathered from a direct measurement or a calculation of an activity or a process. Primary data shows the nature and specific effectiveness of an activity or a process, providing an indication of its specific environmental impact. To give a precise example, in order to calculate the carbon footprint of a building, its energy consumption and characteristics (electricity or gas, renewable or non-renewable etc.) over a specific period of time one would use primary data.

TABLE 2.2 **Examples of emission factors**

Energy (just final combustion)

Natural gas	0.205 kg CO$_2$/ kWh
Green electricity	0 kg CO$_2$/kWh
Electricity (US average)	0.515 kg CO$_2$/kWh
Electricity (EU27 average)	0.460 kg CO$_2$/kWh

Materials or products (life-cycle analysis)

Paper	2 kg CO$_2$/kg of paper
Recycled paper	1.3 kg CO$_2$/kg of paper
Plastic (average)	1.3 kg CO$_2$/kg of plastic
Steel (blast furnace)	3.2 kg CO$_2$/kg
Steel (electric arc furnace)	0.9 kg CO$_2$/kg of steel
Steel (average)	1.7 kg CO$_2$/kg of steel
Aluminium (average)	2.0 kg CO$_2$/kg of aluminium

Transport (just energy)

Bus	0.07 kg CO$_2$/km.passenger
Car (US average)	0.24 kg CO$_2$/km.passenger
Car (EU average)	0.12 kg CO$_2$/km.passenger
Intercity Rail (US EPA)	0.09 kg CO$_2$/km.passenger
Aeroplane (economy)	0.137 kg CO$_2$/km.passenger
Aeroplane (business)	0.398 kg CO$_2$/km.passenger
Diesel	2.66 kg CO$_2$/litre
Petrol	2.42 kg CO$_2$/litre

Sources: Ademe, Defra, IEA, Ecoinvent and US EPA.

When calculating GHG emissions it is often impossible to obtain specific data due to a lack of information or insufficient time to carry out measurements and calculations from direct measurements. Under these circumstances one would use *secondary data*, which is derived from sources other than direct measurements or calculations based on direct measurements. Take the example of establishing the carbon footprint of a building without energy meters. We would have to calculate the GHG emission balance on the basis of an estimate. For example, we might find a reference in the specialist literature indicating that a building constructed during the same period, using the same materials as the building in question, in a comparable climate, requires a total energy consumption of 130 kWh/m^2/year. Assuming this secondary data is correct, we could then advance our study by inputting the floor area of the building. Of course, if we were to replace specific values with average values the level of uncertainty would increase.

The concepts of primary and secondary data can also be applied to emission factors. Imagine that we want to calculate the GHG emission balance for a beer that is distributed in glass bottles. To calculate the packaging emissions in CO_2 equivalent, we need to know the bottle mass, for instance, 300 grams, and the emission factor of the glass, for instance 791 g CO_2/kg, according to the European Container Glass Federation (FEVE, 2012). In this example, the activity data (bottle mass) is primary data that can be obtained simply with a set of high-precision scales. On the other hand, it is more complicated to obtain a specific primary emission factor for the container. It would be necessary to trace all the energy and material flows throughout the entire production chain of the glass container. Someone commissioning a study on GHG emissions for a brand of beer would certainly not have the resources and access to the information necessary to study these flows in the glass sector. The emission factor used will also be a default value, for example the value published by FEVE. In this case it is vital to mention this assumption, as it is scientifically questionable to use an average as a specific value, and FEVE itself advises against using this figure in other contexts. Despite the warning, this type of approximation is necessary for practitioners of carbon accounting. In fact, in this real-life example, most glass suppliers who were contacted hid behind the average data to avoid revealing their specific values.

It would be desirable to require the use of primary data for each product category, in view of the influence that the data has on the overall balance. This could be implemented using a generic life-cycle assessment (LCA) for each product category. The aim is to identify the stages in the life-cycle that make the most significant contribution, and within these stages to determine the key parameters. Practice shows that voluntary carbon-footprint calculations generally use primary data only if it is available to the person commissioning the study, and not necessarily on the basis of the most significant impacts.

Box 2.1   The general principles of carbon accounting

Relevance
The boundaries of GHG emissions accounting and reporting should be defined appropriately. Boundaries are chosen on the basis of the entity under study, the reason for gathering information on GHGs and users' needs.

Completeness
Ideally all emissions sources within the specified organizational and operational boundaries should be reported. In practice, the lack of data or the cost of collecting new data can be a restricting factor. If certain sources are not reported, these omissions should be clearly indicated in the report.

Consistency

People using information about GHGs often want to follow the changes in emissions over time to determine trends and evaluate the carbon efficiency of the organization. It is vital to use the same methods of calculation and presentation. Any change in accounting methodology should be clearly communicated. When presenting the GHG balance it is also important to specify the context and to justify and explain all significant changes. This makes it possible to compare equivalent data.

Transparency

Information is transparent if it ensures that the subjects studied can be well understood in the context of the reporting organization, and if it provides an objective assessment of performance. Using independent external auditing is a good way to increase transparency.

Accuracy

Data accuracy is important for all decision-making. Mediocre reporting systems and inherent uncertainty in the calculation methods applied can compromise the accuracy of the results. Data accuracy can be improved by adhering to the prescribed and tested GHG calculation methods and implementing a solid accounting and reporting system with appropriate internal and external controls.

Source: WRI and WBCSD (2001, 2004).

The concept of boundaries

The question of the area of application is fundamental to carbon accounting. Moreover, it is one of the key elements in initiatives to standardize carbon accounting methodologies. ISO standards distinguish between organizational boundaries and operational boundaries.

The *organizational boundaries* define the geographical limits of the study (organizational entities and sites). The *operational boundaries* (also known as "scope") specify the activities, products and services covered. The Greenhouse Gas Protocol and the ISO standards on carbon accounting recognize three scopes.

Scope 1 includes direct GHG emissions, that is, emissions from sources owned or controlled by the reporting organization. Official inventories (territory-based approach) study emissions only within these boundaries. Direct emissions are mainly the result of the following activities:

- Production of electricity, heat or steam from combustion of fossil fuels;

- Certain physical or chemical processes (e.g., emissions occurring during production of cement, lime, adipic acid, nitric acid and ammonia);

- Transportation of merchandise, waste and people using fossil fuels (e.g., vehicles with internal combustion engines);

- Emissions from the agricultural sector (methane emitted by cattle, the impact of deforestation etc.);

- Fugitive emissions, that is, intentional or accidental emissions due to problems with seals: for example, methane emissions from mines, eHFC emissions from the use of refrigeration and air-conditioning equipment, or methane emissions from the processing and transportation of natural gas.

Scope 2 includes indirect emissions associated with production of electricity, heat or vapour that is imported or bought (i.e., not emitted directly in the territory/site being studied). These emissions associated with generation of imported electricity, heat or steam are a special case of indirect emissions. Consumption of electricity represents one of the main sources of GHG emission reductions for many companies or territorial authorities. They can reduce or rationalize their consumption by investing in technology in order to improve energy efficiency. Companies can also install an efficient cogeneration plant instead of importing more GHG-intensive electricity from the grid. Scope 2 facilitates transparent accounting of these options. If only scope 1 is considered, certain measures, such as the example given concerning the installation of an efficient cogeneration plant, reduce the global climatic impact but have a negative effect on the GHG balance of the entity under scope 1.

Scope 3 covers other indirect emissions that are a consequence of the activities of the reporting company, but arise from sources owned or controlled by another company, such as:

- Employee business travel;

- Transportation of products, materials and waste with vehicles not controlled by the company calculating its GHG emissions;

- Outsourced activities, contract manufacturing, and franchises;

- Emissions from waste generated by the reporting company when the point of GHG emissions occurs at facilities that are owned or controlled by another company;

- Emissions from the use of products and services sold by the company;

- Employees commuting to and from work with private vehicles;

- Production of materials and products bought by the company;

- Purchase of agricultural products;

- Etc.

The latest version of the GHG Protocol defines 15 categories for scope 3 (see Appendix). To illustrate the three scopes, let us imagine an SME specializing in IT hardware maintenance which occupies a building in Manchester, owns five vehicles and consumes electricity for its activities. This company wants to carry out an inventory of its GHG emissions. Which emissions should be included in its accounting and to which scopes should these emissions be assigned?

In our example, the IT company should enter emissions related to its heating system under scope 1. This would include, for example, combustion of 5,000 litres of heating oil per year in a boiler and combustion of fuel for its five vehicles (e.g., 5,000 litres of diesel fuel). If the company uses an air-conditioning system there may be fugitive emissions. Using the emission factors for diesel/heating oil and considering refrigerant losses to be very low (e.g., 100 g per year of R134a, a refrigerant gas with $GWP_{100 \text{ years}}$ of 1,430), emissions under scope 1 are less than 19 tonnes of CO_2-equivalent per year, as shown in the calculation below:

Scope 1 emissions:
(5,000 litres + 5,000 litres) * 2.66 kg CO_2/litre + 0.1 kg
$$* \ 1,430 \text{ kg } CO_2/\text{kg} = 26,743 \text{ kg } CO_2$$

Supposing that the company consumes 20,000 kWh of electricity per year, and taking the average rate of CO_2 for electricity production in the country where the SME is located, we have all the necessary data to calculate scope 2 emissions. According to DEFRA, the average figure for the UK in 2016 was 412 g CO_2/kWh. The emissions in this category are therefore 8.2 tonnes of CO_2 equivalent, as shown in the calculation below:

Scope 2 emissions:
20,000 kWh * 0.412 kg CO_2/kWh = 8,240 kg of CO_2

We note that this emission factor was 90 g CO_2/kWh for France, 430 g CO_2/kWh for Germany, 515 g CO_2/kWh for the US and 640 g CO_2/kWh for Poland. The significant differences between countries can be explained by the energy sources used to generate electricity: mainly nuclear power in France, principally coal in Poland. These emissions factors will lower in most countries with the growing share of renewable energy, particularly in Germany, the UK and several US states where policy-makers are aiming at carbon-neutral grids by 2050.

In our example it would take a significant amount of work to calculate all items under scope 3. It would be necessary to carry out an inventory of all business travel using vehicles not controlled by the company, to identify all purchases of goods and services, to calculate the carbon footprint of commuting etc. It would also be necessary to find appropriate (i.e., the most specific) emission factors for each piece of activity data. Scope 3 could also include indirect emissions due to electricity losses (if the company consumes 20,000 kWh of electricity, it is likely that the supplier had to generate between 21,000 and 22,000 kWh given grid losses).

The idea of scope is sometimes confused with the idea of responsibility. Following this logic, a company would bear more responsibility for scope 1 emissions. However, the reality is often more complex. A company that rents a poorly insulated building may have high scope 1 emissions because it controls these emissions. The opposite could be the case for a company running a just-in-time stock-management system with suppliers located a long way from its base of activities. Its scope 1 emissions might be very low, for example, because of its small warehouse, but it might be responsible for significant scope 3 emissions due to its strategic choices. The most important factor in calculating a carbon footprint in order to reduce global GHG emissions is to identify the most relevant variables for change, whatever the scope. The main advantage of the idea of scope is that it enables the consolidation of emissions from different companies/geographical entities and avoids double counting. Indeed, a company's scope 1 emissions are the only emissions that can be added up between companies without creating double counting. Scope 2 and 3 emissions must be scope 1 emissions for another entity. For example, scope 2 emissions in France are often scope 1 emissions for EDF or Engie, two of the main suppliers of electricity.

The notion of operational boundaries can also be applied to products. For example, a European regulation obliges car makers to publish the CO_2 produced by the vehicles they sell; however this figure covers only direct emissions resulting from the vehicle user's fuel consumption (scope 1). The car makers do not have to publish emissions from other scopes, or embodied emissions (also referred to as embedded or "grey" emissions), linked to manufacture, transportation and storage of the product. Given the high number of sub-contractors involved, the number of options available for a vehicle and the fact that the same model is often assembled at different assembly plants, it is methodologically impossible to calculate embodied emissions from manufacture. A car that emits 140 g of CO_2 per km and travels approximately 200,000 km in its service life produces combustion-based emissions of around 28 tonnes of CO_2. If we carry out a quick analysis of emissions linked to manufacturing, the main elements of the car (1.2 tonnes of steel, 100 kg of various plastics, 5 tyres, various electronic components, logistics, the energy required for assembly, end-of-life waste processing etc.) we can add approximately 7 tonnes of CO_2 for embodied emissions. In this example embodied emissions represent only 20% of total emissions for the entire life-cycle (7 tonnes out of a total of 35 tonnes emitted from manufacture to the end of the vehicle's life). Faced with this assessment, public authorities and consumers focus their attention on the direct emissions from vehicle use.

A similar assessment could be carried out for the building sector, where life-spans are longer and emissions linked to construction (concrete, steel, glass, transportation of materials, energy used on-site) often account for less than 10% of total emissions over a life-cycle. Indeed, buildings often consume more energy while they are being used (heating, lighting, ventilation etc.).

However, in the case of transport and buildings the situation is changing quickly, and in future the opposite might apply. Indeed, voluntary commitments followed by binding regulations have reduced the fuel consumption of vehicles, and increasingly strict regulations have been adopted to govern energy consumption in buildings. As a result GHGs emitted during manufacture could become the main source of emissions over entire life-cycles. This trend could pick up speed with the rise of electric vehicles and passive houses where energy needs are sufficiently low to render heating systems redundant.

3

Official inventories: the territory-based approach

Official inventories, which were developed systematically after the UNFCCC came into force, analyse emissions taking place in a particular territory. It is worth noting at the outset that before the Paris agreement, reporting obligations and frequency vary depending on whether a country is included in Annex I or not. Developed countries (Annex I) are obliged to publish their inventory annually, whereas emerging countries such as China or India publish their inventory every six or seven years on average. Once the Paris agreement comes into effect, all countries will have to submit these emission inventories at least every two years. The IPCC played an important role in setting up standardized methodologies for carrying out these inventories.

The territory-based approach has a more robust methodology than the "footprint" approach, and it has also been chosen for emissions inventories for companies participating in the European carbon market. Voluntary approaches are also generally seen to prefer "footprint" approaches with broader scopes (see Box 3.1, Covenant of Mayors and Chapter 4).

The national inventory report

Official inventories of Parties to the UNFCCC are carried out on the basis of emission registers. This approach takes into account only emissions originating in the territory subject to study. Unlike "footprint" approaches, this analysis enables one to add up different inventories with no risk of double counting and to produce maps comparing emission origins. The official report, known as the *National Inventory Report*, provides a complete description of GHG emissions under the UNFCCC, information about the activities responsible for emissions or absorption and a description of the methods used for estimation.

The GHGs estimated in national inventories are carbon dioxide (CO_2), methane (CH_4), nitrous oxide (N_2O), sulphur hexafluoride (SF_6), perfluorocarbons (PFC) and hydrofluorocarbons (HFC). Presentation of these inventory items is harmonized at the international level.

Emissions or absorption are generally calculated or estimated using mass balances and emission factors. The experts responsible for producing the inventory analyse and check the methods, activity data and emission factors to ensure that the estimates for emissions and absorption are of high quality. These experts are also responsible for managing quality assurance and archiving systems, and for trend analysis. Quality assurance and quality control procedures must ensure that the general principles of carbon accounting are respected (see Box 2.1).

Expert Review Teams (ERTs) made up of four to 12 individual experts check that the inventories are complete, reliable and in compliance with directives. Each team has two lead reviewers, one from an Annex I Party and the other from a non-Annex I Party. The annual inventory reviews are generally carried out at the office of the UNFCCC Secretariat in Bonn, Germany. However, each Annex I country has been the subject of at least one in-country visit during its period of commitment. Following the example of company financial audits, if problems or inconsistencies are observed, the ERT can recommend that the data be adjusted to ensure that emissions for a given year are not underestimated. The Compliance Committee intervenes if there is a disagreement between a Party and the ERT.

Annex I Parties must also supply national communications on the actions they are taking to implement the Kyoto Protocol. Each communication that is submitted is also subjected to a detailed review by the ERT. For non-Annex I Parties, the national communication is the only obligatory document, and it also includes an inventory of emissions. However, as previously mentioned, the reporting period is less strict. The largest emitter in the world, China, published its latest and second national communication in November 2012, showing an inventory of its emissions for 2005.

The IPCC guidelines

The 2006 IPCC guidelines for national GHG inventories propose methodologies to estimate national inventories of anthropogenic GHG emissions by sources and absorption by sinks (IPCC, 2006). The 2006 IPCC guidelines were drawn up at the invitation of the Parties to the UNFCCC.

The 2006 guidelines consist of five volumes. The first provides general guidance and the remainder cover specific subject areas (energy, industrial processes, agriculture and waste). The 2006 guidelines were produced by consolidating and updating the 1996 guidelines, the "IPCC Good Practice Guidance and Uncertainty Management in National Greenhouse Gas Inventories" (IPCC, 2000a) and the "IPCC Good Practice Guidance for Land Use, Land-Use Change and Forestry" (IPCC, 2000b).

The 2006 guidelines offer recommendations for estimation methods based on three levels of detail (*tier approach*), from level 1 (the default method) to level 3 (the most detailed method). If used correctly, each level can produce unbiased estimates, with gradually increasing accuracy from level 1 to level 3. The fact that there are different levels enables the bodies responsible for inventories to use methods that suit the available resources and concentrate their efforts on the emission and absorption categories that make the most significant contribution to emission trends or totals in a given country.

The guidelines provide presentation tables and worksheets for level 1 methods. The emission factors included in these guidelines are often used as the default value for establishing national inventories, and they are also a reference for inventories in industry and voluntary inventories.

These guidelines are used by developed countries. For instance, the US EPA used them for the development of the US emissions inventory.

Methodologies within the EU ETS

The EU has set up a European Union Emissions Trading System (EU ETS) for GHG quotas. This system will be described in more detail in Chapter 4. This section deals with the guidelines and the principles of carbon accounting developed within the framework of the system. Following the example of the UNFCCC, a territory-based approach was chosen, although companies have to calculate and limit only their direct GHG emissions. In other words, the system covers only scope 1 emissions.

Each plant covered by the EU ETS must obtain an emission permit, that is, an authorization from the authorities responsible for GHGs controlled by the Kyoto Protocol. One of the conditions for obtaining this permit is that operators must be able to monitor and report the emissions caused at their plant. A permit is not the same as a quota: the permit describes the monitoring and reporting obligations for a plant, whereas a quota or allowance is a unit of exchange in the system. The rules for allocation of quotas will be described in the last chapter. Operators must declare emissions falling within the framework of the EU ETS at the end of each calendar year. Reports must be audited by an independent body in accordance with the criteria defined in the legislation on the EU ETS, then published.

During the first compliance cycle of the EU ETS in 2005 the operators, auditors and competent authorities of the member states gained their first experience of monitoring and reporting thanks to Commission Decision 2004/156/CE (EU, 2004a). This decision includes the basic principles of carbon accounting for industrial plants. It specifies that plant emissions should be monitored using calculations or measurements. The following formula is used for the calculations:

$$Emissions = Activity\ data \times Emission\ factor \times Oxidation\ factor$$

The last factor, representing the proportion of carbon that is not oxidized during the chemical reaction, is not necessary if the emission factor already takes into account the fact that a proportion of the carbon is not oxidized. In practice the oxidation factor is often between 98% and 100%. The guidelines state that the default emission factors are acceptable for all fuels except non-commercial fuels (combustible waste such as tyres and gases from industrial processes). Seam-specific emission factors were produced for coal, and EU-specific or producer-country-specific default factors were produced for natural gas. The IPCC default values are accepted for refinery products. The emission factor for biomass is zero, which clearly tends to encourage companies within the system to use more biomass instead of fossil fuels. The accounting does not differentiate between sources of biomass, sometimes leading to perverse effects, for example energy that is produced using palm oil from new plantations that were created by cutting down primary forests. If an operator chooses to measure emissions at chimney outlets, the measurement methods must be standardized, recognized and corroborated by an emissions calculation.

These guidelines specify and emphasize the important principles of carbon accounting inspired by the principles in the Greenhouse Gas Protocol, also adding to them: completeness, consistency, transparency, trueness, faithfulness and cost effectiveness. The last principle attempts to balance the positive effects of increased accuracy with the additional costs caused.

Decision 2004/156/CE was replaced in 2007 by Decision 2007/589/CE (EU, 2007). The changes took into account recommendations on GHG monitoring from the IPCC and the International Organization for Standardization (ISO).

However, experience has shown that there are divergences between monitoring, reporting and auditing practices between member states. To increase the system's credibility, the guidelines were replaced by harmonized regulations No 525/2013 in May 2013. The new decision notably includes all changes carried out following integration of aviation activities.

Box 3.1   The (Global) Covenant of Mayors

The Covenant of Mayors is a movement that brings together local and regional authorities on a voluntary basis to improve energy efficiency and increase the use of renewable energy in their territories. The first signatories to this Covenant were committed to meeting or exceeding the EU objective of reducing CO_2 emissions compared to 1990 by 20% by 2020. Since October 2015, the new signatories now pledge to reduce CO_2 emissions by at least 40% by 2030 and to adopt an integrated approach to tackling mitigation and adaptation to climate change. The signatory towns must immediately carry out an audit of their CO_2 emissions to achieve this objective. By August 2016, this movement had 6,890 signatories, representing 213 million Europeans.

The Covenant of Mayors officially merged with the Compact of Mayors on 22 June 2016 in Brussels, Belgium. The Compact was launched in 2014 by UN Secretary-General Ban Ki-moon and former New York City Mayor Michael Bloomberg, the UN Special Envoy for Cities and Climate Change. The Compact represents a common effort from global city networks C40 Cities Climate Leadership Group (C40), ICLEI, and United Cities and Local Governments (UCLG), as well as UN-Habitat, to unite against climate change (before the merger, 428 global cities had committed to the Compact of Mayors).

The newly created Global Covenant of Mayors for Climate and Energy unites more than 7,100 cities in 199 countries across six continents in the shared goal of fighting climate change through coordinated local climate action. The initiative represents more than 600 million residents, or more than 8% of the world's population.

Although voluntary in nature, it is interesting because it studies emission registers at a sufficiently granular level to enable reflection on town planning and regional development. However, there are clear disparities in the quality of participant inventories and the level of ambition in measures to improve energy efficiency or increase the proportion of renewables. In particular, the reference point is often very vague, as municipalities are frequently unable to estimate their historical emissions in 1990, choosing instead a more recent reference date that displays higher emissions due to the increase in energy consumption in the building and transport sectors, the two main inventory items in urban areas. Another problem is that no specific calculation methodology is required, and certain towns do not use a territory-based approach, but rather analyse their territorial carbon footprint by carrying out an emissions inventory covering all three scopes. The confusion between territory-based approaches and "footprint" approaches is symptomatic of voluntary approaches. A further criticism, also linked to the voluntary nature of the exercise, is that those initiating the approach may be looking for a quick communications payoff, leading to a risk of greenwashing.

For more information see: http://www.covenantofmayors.eu/

EPA's Greenhouse Gas Reporting Program (GHGRP)

As directed by Congress, the EPA's Greenhouse Gas Reporting Program (GHGRP) collects annual GHG information from the top emitting sectors of the US economy. The GHGRP is the only dataset containing facility-level GHG emissions data from large industrial sources across the United States. With five years of reporting for most sectors, GHGRP data are providing important new insights on industrial emissions. They illustrate variation in emissions across facilities within an industry, variation in industrial emissions across geographic areas, and variation in emissions over time at the sector and facility level. The EPA is using this facility-level data to improve estimates of national greenhouse gas emissions, including using it to improve the US Greenhouse Gas Inventory. The data are also being used to inform regulatory actions and voluntary emission-reduction efforts.

In 2016, over 8,000 facilities and suppliers reported to the GHGRP. Their efee emissions totalled 3.20 billion metric tonnes CO_2-equivalent, or about half of all US GHGs. The covered facilities are those that emit 25,000 metric tons or more per year.

Also covered were 957 suppliers of products that generate GHG emissions when released, combusted or oxidized, and 92 facilities that inject CO_2 underground for geologic sequestration or other purposes.

While the GHGRP prescribes methodologies that must be used to determine GHG emissions from each source category, reporting entities generally have the flexibility to choose among several methods when calculating GHG emissions. The decision of which method to use may be influenced by existing environmental-monitoring systems and other factors. This flexibility is also illustrated by the fact that reporters can change emission-calculation methods from year to year or even within the same year, as long as they meet the requirements needed to use those methods.

The reporting program was envisioned as a key part of a nascent federal climate policy, but that policy fizzled while the reporting program was developed. Nevertheless, the value of the reporting data should not be underestimated. The data are likely to support further EPA regulations under the Clean Air Act, such as allowing modified performance standards for permitted stationary sources.

Global Protocol for Community Scale Greenhouse Gas Emission Inventories (GPC)

In December 2014, World Resources Institute, C40 Cities, and ICLEI revealed the Global Protocol for Community Scale Greenhouse Gas Emission Inventories (GPC), identified as the first set of standardized global rules for cities to measure and publicly report their carbon pollution emissions. This standard is the baseline on which Compact of Mayors was founded (see box).

4

Voluntary inventories: the "footprint" approach

Introduction to carbon footprints

A carbon footprint is defined as a measure of total GHG emissions caused directly or indirectly by a person, organization, product or service. An *organizational carbon footprint* reports all emissions that may take place in an organization, including energy used in buildings, industrial processes and company vehicles. Depending on the scope of the study, indirect emissions linked to consumption of goods and services or grey energy from materials may or may not be included. To take these into account it is necessary to estimate the carbon impact of the products purchased by the organization – a procedure that is often imprecise and complicated to implement. A *carbon footprint for products or services* takes into account emissions during the entire life of a product, from extraction of the raw materials, production and use to the end of the product's life, including recycling, treatment and disposal. Although somewhat confusing, it is also possible to carry out a *territorial carbon footprint* taking into account the emissions for goods consumed in the territory. Indeed, the main difference between a carbon-footprint approach and a territory-based approach is that the former is "consumption-based", while the latter is "production-based" (Kander *et al.*, 2015).

An increasing number of organizations are calculating their carbon footprints for various reasons. Some organizations do so in order to "green" their image or as part of their corporate social responsibility policies (Capron and Quairel-Lanoizelée, 2010). Others do so to prepare for binding regulations that may be introduced in the future or to identify ways to improve their organization, which they may be able to integrate into their environmental management system (ISO 14001 or EMAS), for example. Carbon footprints are also sometimes calculated because of demand from customers or other stakeholders (particularly in the case of financial institutions through the *Carbon Disclosure Project* – see Box 4.1).

The increased use of the "footprint" approach by public authorities reflects a desire to improve the clarity of the territory-based approach, which does not show the climatic impact of economic choices and activities carried out in a territory or an organization. Indeed, the increased mobility of goods, people and capital mean it is now vitally important to take into account "carbon leakage" when defining truly effective climate objectives. Studies using data on bilateral trade and CO_2 emissions show the limits of an approach focusing exclusively on direct emissions. In France, the Observation and Statistics Directorate (SOeS) in the General Commission for Sustainable Development (CGDD) carried out an exercise of this type revealing the level of hidden carbon in imported goods (CGDD-SOeS, 2010). The study demonstrated that OECD countries have a significant carbon deficit. The OECD estimates that if the GHG emissions of France's imports were included and the emissions linked to exports were subtracted,

French emissions in 2000 would have been 35% higher than the figure given by the official inventory approach (Nakano *et al.*, 2009). Continued relocation of industries to countries where labour costs are lower has increased the phenomenon of carbon leakage. Other more recent studies highlight this marked trend in OECD countries (Glen *et al.*, 2011).

Although it is imprecise, analysis carried out with extended boundaries can enable us to calculate the climate benefit of innovative solutions. For example, introducing concentrated washing powders has been an effective way for most manufacturers to reduce GHG emissions from logistics and packaging, two sources of indirect emissions. To reduce the carbon impact of their products, these manufacturers principally had to work with their suppliers (logistics providers, manufacturers of defoaming agent and packaging, electricity suppliers etc.).

Nevertheless, as we will see, using an approach based on footprints rather than direct emissions poses problems for methodology and data access. This costly approach, which is little suited to determining reference values or emissions standards, needs to be further developed before it is incorporated into binding legal instruments. At the moment, carbon footprint calculations are generally voluntary. In the case of exceptions, the boundaries are generally limited and the sources are clearly defined, for example Article 75 of the Grenelle Act in France obliges large companies and local authorities to calculate their carbon footprints for scopes 1 and 2.

Organizational carbon footprints

The GHG Protocol

The *GHG Protocol* was instituted in 1998 by a partnership between the World Resources Institute (WRI) and the World Business Council for Sustainable Development (WBCSD). This multilateral partnership of companies, non-governmental organizations (NGOs) and governments proposed the first accounting framework for company GHG emissions standards, programmes and inventories, and it serves as a benchmark. The first edition of *The Greenhouse Gas Protocol. A Corporate Accounting and Reporting Standard* was published in 2001 (WRI and WBCSD, 2001). It was updated in 2004 and many supplementary instruments and reference documents have been produced. Application of the framework proposed by this standard is still voluntary. Regulations are generally more likely to refer to standards from recognized standards organizations, particularly the ISO.

The protocol deals with all GHGs covered by Kyoto. If companies define their own organizational boundaries, they can choose to organize their GHG emissions inventory using criteria linked to operational control (e.g., including franchises) or financial control (e.g., including emissions by companies part-owned by the organization). Companies are encouraged to do their accounts and publish emissions for at least scopes 1 and 2. In this voluntary exercise, scope 3 emissions do not have to be published.

The Protocol insists on the importance of having a quality management system and gives five principles of carbon accounting to ensure emissions are portrayed accurately (see Box 2.1, The general principles of carbon accounting).

Two additional voluntary standards were published at the end of 2011: The *Product Life Cycle Accounting and Reporting Standard* (see "The *GHG Protocol Product Life Cycle Accounting and*

Reporting Standard") and the *Corporate Value Chain (Scope 3) Accounting and Reporting Standard*. The second of these standards provides a specific interpretation of scope 3, identifying 15 categories of emissions, both upstream and downstream from companies' direct emissions and emissions linked to companies' consumption of electricity and heat (WRI and WBCSD, 2011a). This standard was intended to help organizations gain a better understanding of emissions over the entire value chain.

In 2014, the *Mitigation Goal Standard* was released. It provides a standardized approach for assessing the progress of national and sub-national GHG-reduction goals and can help governments set emission-reduction targets, meet domestic and international emissions-reporting obligations to groups like the UNFCCC and ensure that efforts to reduce emissions are achieving their intended results (WRI and WBCSD, 2014).

Box 4.1   CDP

CDP (formerly referred to as the **Carbon Disclosure Project**) is an independent non-profit organization which has the largest database in the world of primary information from companies concerning their GHG emissions, their carbon-reduction strategies and their use of natural resources. Launched in 2001 by a consortium of institutional investors, the CDP requests information every year from the largest companies in the world. In 2016, the 827 institutional investors that participated in the initiative had combined assets of more than $100 trillion. Information is gathered in four broad subject areas:

1. Opinion and direction regarding the risks and opportunities that the company faces when combating climate change;

2. Carbon accounting;

3. Strategy and efforts used to reduce emissions, mitigate risks and employ opportunities;

4. Climate-change management in the company.

The first request for information was sent in 2003, and 235 companies responded. There are now more than 20 times as many companies participating, as the number has grown to more than 5,000 companies. These companies are located in more than 60 countries and are involved in all sectors of economic activity. The response rate has steadily increased, and in 2016 it was over 89% for the largest 500 companies in the world. More recently, CDP has extended its activities to include gathering information on climatic impact throughout the supply chain (*Supply Chain Program*), climate change relating to municipalities (**Cities**), water consumption by companies (*Water Program*), and companies' exposure to deforestation risks (*Forest Program*).

As it is a voluntary initiative, companies are free to choose the organizational boundaries and the scope for their communications. For example, many companies who operate in southern countries decide not to report their emissions. There is also no institutional control of the data they communicate. However, this first step towards more transparency around corporate carbon footprints could enable further controls by stakeholders including the authorities.

For more information see: www.cdp.net.

The Environmental Reporting Guidelines: including mandatory GHG emissions reporting guidance (United Kingdom)

Initially referred to as the Guidelines for Company Reporting on Greenhouse Gas Emissions, this was the first carbon-accounting method intended for companies when it was published by the Department of the Environment, Transport and the Regions in the United Kingdom in 1999. The Department for Environment, Food and Rural Affairs (DEFRA) and the former Department of Energy and Climate Change (DECC) publish an annual quality methodology guide including specific emission factors for a British context, making a useful addition to the range of methods available.

The latest version of the guidelines is designed to help companies comply with the GHG reporting regulation, a requirement from the Climate Change Act 2008, as well as aid all organizations with voluntary reporting on a range of environmental matters, including voluntary GHG reporting and through the use of key performance indicators.

This guidance includes changes which took effect from 1 October 2013, requiring all UK-quoted companies to report on their GHG emissions as part of their annual directors report. That requirement affects all UK-incorporated companies listed on the main market of the London Stock Exchange, a European Economic Area market or whose shares are traded on the New York Stock Exchange or NASDAQ. The UK government encourages but does not require all other companies to report similarly.

This reference base initially only covered direct emissions and indirect emissions from electricity and heat use (scope 2), before it was extended to include certain other indirect sources for companies (business travels, logistics and waste management). More than 15 years after the first release and thanks to numerous updates, it remains a trusted source. The yearly publication of "conversion factors" are for instance used by many companies that report on a voluntary basis, including in the United States where the EPA's Center for Corporate Climate Leadership recommends Britain's guidelines on calculating air-travels emissions.

The Bilan Carbone®

The Bilan Carbone® (carbon balance) is the most widely used carbon-accounting methodology in France. The idea was conceived by Jean-Marc Jancovici, an engineer who did much to raise public awareness of climate issues and ADEME. The Bilan Carbone®, along with its underlying emission factors, provides a comprehensive methodological foundation. Between 2004 and 2012, more than 4,000 people were trained in the use of this carbon-diagnostics tool.

This voluntary method distinguishes three "boundaries", and in the initial versions they were defined in a different way to the three scopes in the ISO standards and the Greenhouse Gas Protocol. Version 7 of the Bilan Carbone® method resolved this issue.

The Bilan Carbone® is based on emission flows, and not on the particular location of emissions. This means that all emissions that are necessary for an activity are taken into account, and the location of the emission is not important. The method enables users to decide how far to take their investigations to gain a sufficiently clear picture of the emissions associated with their activities. The Bilan Carbone® makes it possible to identify opportunities for reductions with suppliers and clients.

One of the principles of the method is to equate GHG emissions that take place within an organization (for which it is in some ways directly responsible, in legal or territorial terms) and emissions that take place outside the organization, but which result from processes required for the activities of the organization in its current form.

As choices are left open to the user and the extent of application under scope 3 is unclear, the Bilan Carbone® provides results only to an order of magnitude, and it does not enable comparisons between operators or with reference values. It is therefore a tool for internal use only, and any publication of carbon-balance results is risky as the area of investigation is so vague.

It is worth noting that in addition to a version for companies, the ADEME has developed two modules specifically for public bodies: the "territories" module, for calculating territorial carbon footprints, and the "assets and services" module for organizational carbon footprints of public authorities. These tools are often used by French signatories to the Covenant of Mayors (see Box 3.1) and by some municipalities outside France (certain versions of the tool have been translated into English).

At the end of 2011 ADEME stopped managing this method, and a private body, the Association Bilan Carbone (ABC – the Carbon Balance Association), took over. The aim of this association, which brings together local authorities, companies, experts, institutions and consultants, is to unite private and public actors who use the method and want to develop and promote the Bilan Carbone®. To use the latest version of the tool and have access to updates, it is necessary to be a member of the ABC and to pay an annual licence fee. Training in the method is provided by the Institut de Formation Carbone (IFC – Carbon Training Institute). The ADEME now concentrates on three tasks relating to carbon accounting: creating prescribed methods, developing a Base Carbone® (a carbon reference base for emission factors in France, available at www.basecarbone.fr) and promoting voluntary GHG balances (using Bilan Carbone® or other comparable methods).

Will it be compulsory for organizations to report their carbon footprint?

As voluntary carbon-footprint-calculation processes have developed, authorities have made increasing efforts to regulate them or make them obligatory, using varying boundaries.

In the United Kingdom, many companies already publish their GHG emissions as part of climate-change agreements (voluntary mechanism) or the *Carbon Reduction Commitment*, an obligatory emissions-trading system linked to energy consumption, launched in April 2010. This system requires about 5,000 organizations to monitor and publish their GHG emissions, and it requires 15,000 others to publish their electricity consumption. Reporting is not restricted to just direct emissions, as consumption of electricity is also included. However, the UK government announced in 2016 that the CRC energy efficiency scheme will be abolished following the 2018 – 19 compliance year.

In the United States, the EPA published a directive for publication of GHGs in 2009, which obliged facilities emitting more than 25,000 tonnes of CO_2 to publish their emissions from combustion or industrial processes (GHGRP). The original element of this system was that it requires suppliers of fuels and refrigerant gases to measure the impact of their fuels and industrial gases on the final consumer in CO_2 equivalent (indirect emissions on the final consumer, measured *upstream*). Thanks to this system, the United States has an inventory carried out by the private sector which includes most emissions linked to transport, heating and air conditioning in the service sector and homes.

Although voluntary, the Climate Registry is an influential North American non-profit organization that records and tracks the verified GHG emissions of businesses, municipalities and other organizations. The Climate Registry's Board of Directors is quite unique as it is made up of 31 states of the United States, 13 provinces/territories of Canada, six states of Mexico, and three Native Sovereign Nations. The Registry, launched on May 8, 2007, is modelled on the former California Climate Action Registry and is headquartered in Los Angeles, California.

In France the Grenelle Act 2 (Article 75) made GHG emission calculation obligatory by 31 December 2012 at the latest, for companies with more than 500 employees (or with more than 250 employees in overseas territories), public bodies with more than 250 staff, territorial authorities with more than 50,000 inhabitants, and the state. A decree specifies the content. The emission balance provides "an evaluation of the volume of GHG emissions produced by activities carried out by a legal entity in national territory over the course of a year". It defines two scopes about which the balance must provide information: direct emissions and indirect emissions produced by using electricity, heat or steam required for activities carried out by the legal entity. The decree makes explicit reference to the standard ISO 14064-1. Reporting on scope 3 is still optional.

Reporting obligations in other countries (EU, Japan, South Korea, Australia etc.) are generally limited to direct emissions following the territory-based approach.

Carbon footprints for products

Two factors explain the development of carbon footprints for products and services. Firstly, to calculate an organization's carbon footprint using expanded operational boundaries (scope 3) it is necessary to carry out or update an analysis of a larger number of carbon footprints for products or services. To carry out a precise inventory of companies' indirect emissions, one needs carbon footprints for all significant purchases and investments. Secondly, analysis at a product level enables one to establish reference values and compare the climatic impact of goods and services. This information is useful for implementing new environmental policies, raising awareness among consumers and launching ecodesign approaches. Product carbon footprints carried out using the same assumptions can also be used to compare the climatic impact of products with equivalent functions (see Box 4.2, Choosing a functional unit) and inform choices made by consumers and authorities.

Life-cycle assessment of products and services

Life-cycle assessment (LCA) involves making an inventory and evaluating the flow of materials and emissions, and the impacts associated with a product or service throughout its life-cycle (from design to waste processing). Recent developments in the area of carbon accounting applied to products and services have been directly inspired by the LCA methodology. Indeed, the carbon-footprint calculation for a product involves carrying out an LCA, but limiting the analysis to a single impact category, namely the anthropogenic greenhouse effect.

During the 1970s, LCAs were mainly used to estimate cumulative consumption of energy and raw materials. They were combined with input-output economic models to calculate the emissions and economic costs of products over their entire life-cycle. In the 1980s, when the oil crisis abated, examination of the life-cycle of products was mostly of interest to regulators. LCAs got their second wind when the regulatory framework developed from a policy of nuisance management to an integrated policy for prevention and environmental optimization. In the early 1990s, LCAs were used for communication and marketing purposes. At the same time, the lack of transparency for certain key parameters, the need for assumptions, data availability and the element of subjectivity and interpretation led to inappropriate commercial claims and reduced stakeholders' confidence in LCAs.

Carrying out an LCA is an iterative process, meaning that the first stages can be adapted on the basis of the results of the later stages. For example, if it is difficult to obtain data for the inventory, one can reframe the objectives and the scope of the study. When a very significant impact is identified, it may be wise to refine the data collection that has given rise to this result.

If the analysis can be limited to a single impact, for example in the case of carbon accounting, one must ensure that the impact retained is relevant. Thus, although it may be interesting to carry out GHG emissions balances for bio-fuels in order to compare them with traditional fuels, this type of analysis may not be relevant when comparing two nuclear reactors as the most significant environmental impact is not from GHG emissions. This type of single-criterion GHG emissions analysis should not in any case mask other environmental or social impacts caused by a product.

Because LCAs, thus far, have been carried out on a voluntary basis, we must bear in mind that the results depend on the quality of the information used by the bodies carrying out these LCAs and their good faith. The ISO attempts to limit the risk of manifest error or manipulation of information, and it obliges companies using the standard ISO 14040 as a frame of reference to complete a peer verification of their results (by other academic researchers or specialized consultants) before publication.

The stages of an LCA

An LCA is carried out in four stages, which are described in detail below.

Stage 1: Definition of the goal and scope of the study. The aim is to specify the target audience and the reasons for the study. The reason for the study could be to identify the main impacts of a product, improve an existing product, choose one product over another, produce an innovative communication plan, choose a government environment policy, establish a strategic plan etc.

FIGURE 4.1 **The stages of an LCA**

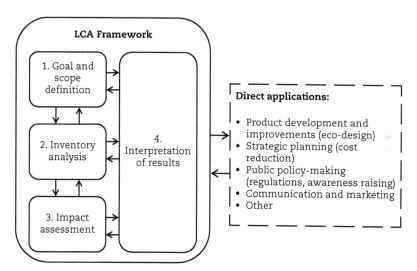

Source: Adapted from ISO 14040 (ISO, 2006a).

It is vital to provide a precise definition of a product's function so that results can be compared (see Box 4.2). A good definition of the function also makes it possible to correctly determine the boundaries of the system to be studied.

Once the function and the attributes have been clearly defined, it is necessary to determine the limits of the system to be studied, which will reveal the requirements of the function. The system to be studied is then broken down into elementary processes (extraction of raw materials, transportation, processing etc.).

Box 4.2   Choosing a functional unit (for example flooring)

What is the function of lino flooring? To protect and decorate the floor. Once one has established the function, it is important to define the *functional unit*. A functional unit should be precise, measurable and additive. In the case of the function mentioned above, the functional unit could be covering 10 square metres of floor for 20 years.

These definitions of function and functional unit are sufficiently open to compare one type of lino with other types of lino or with parquet or tiled flooring, which have an identical role. The choice of function is important. For example, comparing the LCAs (or the carbon footprints, if we restrict the investigation to climate impact) of two different objects is pointless if the function has not been specified over time. It is possible that a durable, high-quality object has a more significant environmental impact during production. However, if we consider its entire lifetime, it may have a more favourable environmental balance. For food products the definition of function is often tricky. For example, does a litre of a fizzy drink have the same function as a litre of orange juice?

To choose the functional unit one needs to identify the **key parameters**, that is, the parameters that allow one to find the defined functional unit for a given product. Service life, the number of possible uses and efficiency are examples of parameters. For flooring the key parameters might be service life and the quantity of maintenance products required. Including maintenance within the boundaries can significantly change the results. The most subjective parameters can come under the definition of function. For example, does lino really have the same function as an inlaid wood floor? An expert carpenter would probably question our choice of functional unit, as an inlaid wood floor is a unique work, with a function that goes beyond simply covering a floor.

Stage 2: Analysis of the life-cycle inventory. The analysis stage of the life-cycle inventory (LCI) involves taking an inventory of all the flows. Two types of flow are identified during a life-cycle analysis:

- economic flows, that is, flows of materials, energy and services between the elementary processes and the elementary systems;
- elementary flows, that is, flows into the biosphere (raw materials, waste returned to the environment and emissions).

This stage is often the longest and most complicated because of the amount of data to collect. Data collection requires surveys on the ground, bibliographic research and reference to previous studies. Some professional federations have assembled data on the environmental impacts of materials in the upstream section of their life-cycle and made this information available to their clients so they can include it in their own LCAs.

Stage 3: Life-cycle impact assessment. The life-cycle impact assessment (LCIA) involves converting an inventory of flows into a series of clearly identifiable impacts. The set of flows in and out are aggregated in the impact categories to provide category indicators. There are various methods for carrying out this type of assessment. On the basis of the activities studied, it is generally possible to distinguish primary effects, such as CO_2 emissions, and secondary effects, such as an increased concentration of GHGs. This increase causes a temperature increase, which can lead to rising sea levels and trigger floods, which in turn have a social and economic impact.

Stage 4: Interpretation. The interpretation aims to draw conclusions and specify the limits of the analysis that has been carried out. During an LCA, the process used is often as important as the final result. This process must therefore be open and comprehensible so that the reader is able to judge the value of the analysis carried out. The interpretation should also clearly establish the limits of the study.

Taking into account an increasing number of criteria in LCAs has raised questions and concerns, and it could lead to people using over-simplifications in order to abuse the system (ADEME, 2005). For example, it could be difficult to decide which is worse, the anthropogenic greenhouse effect or nuclear waste. Some consultants have looked at the different results obtained with a view to obtaining a single score, and software has also been used to examine this question. However, the practice is rejected by ISO standards as there is no consensus on the coefficients that would allow addition by applying different weighting for impacts on nature (greenhouse effect, acidification, waste etc.).

Evaluation of impacts through *monetization* is often used during the interpretation stage. This involves an economic assessment of environmental damage (see Chapter 5 on climate damage). This method is clearly useful for raising decision-makers' awareness of environmental impacts. However, it is currently difficult to put a figure on the external costs. In addition, these costs are often estimated using different methodologies for each impact (cost of damage repair, costs of environmental remediation etc.). As each method has its own uncertainties and biases, the results of different impacts cannot be considered homogeneous nor directly comparable.

Allocation of emissions among co-products

Let us suppose we want to calculate the carbon footprint of a litre of milk. How can we allocate the methane emissions from the cow between milk and meat production? The same difficulty arises with a host of other products. For example, diesel is produced from refining oil, which is used to produce several fuels (diesel, but also petrol and natural gas) and other co-products, such as asphalt.

Various approaches are possible. The simplest solution is to avoid having to allocate the emissions of a process and dividing the multi-function process into single-function processes. Although this solution is simple it cannot be applied yet. In our example, one could theoretically divide the life of the cow and the associated emissions into several sections corresponding to the function of the animal. Until the birth of its first calf, the function of the cow is to produce meat. From its first calf, the cow starts its career as a dairy cow, and direct methane emissions or indirect emissions linked to its food or heating the stable are allocated to milk production. Emissions from the last months of the cow's life, which are dedicated to fattening up, are allocated to meat production.

One could also avoid the need to choose an allocation key for distribution by using an approach that extends the system. This method is used when a process creates a co-product that can be produced using a single-function process. In this case, we extend the boundaries of the system to include this single-function process, and then we subtract its impacts from the multi-function process. In this example, the alternative to a burger from a dairy cow is a burger from a beef cow. The amount attributed to the milk is the total emissions for the "dairy cow" system minus emissions from the "beef cow" system. This approach forces one to carry out a more complex analysis, taking into account an extra process.

The last method available for resolving the problems of multi-function processes is allocation. This involves defining an allocation key for the flows of emissions, reflecting each co-product's level of responsibility on the basis of an "objective" parameter. This parameter could be mass, for example. The emissions are then allocated to each sub-product on the basis of the sub-product mass compared to the total mass of products derived from the process. This means it is possible to carry out the analysis by taking into account only a fraction of the emissions. One could also use the monetary value of the co-products as an allocation factor. Although economic allocation may be a better reflection of the choices made by producers (i.e., the reason a process is carried out), it is often criticized for its dependence on price fluctuations. With this approach, CO_2 emissions for milk rise if the price of milk goes up more than the price of beef, even though there has been no change in the milk production method.

It is often necessary to choose the allocation method when carrying out an LCA, and it has been criticized for the level of variation that this choice causes. A literature review of different LCAs for milk shows that the carbon footprint found for milk is higher using mass allocation (95% of emissions from cows are allocated to milk and 5% to meat) than with monetary allocation (85% of emissions from cows are allocated to milk and 15% to meat). If emissions are not allocated by extending the system or dividing the life of the cow into functional segments, the proportional responsibility attributed to milk production as opposed to meat production is also reduced.

Publicly Available Specification 2050 (PAS 2050)

PAS 2050 was the first standardized methodology for calculating the carbon footprint of products or services (BSI, 2008). It was published in 2008 by the BSI (British Standards Institution) at the request of

DEFRA and the Carbon Trust, a charity set up by the British government to help companies reduce their GHG emissions. This methodology enables the conversion of GHG emission flows into CO_2 equivalent for activities in the industrial and service sectors, by product or service type. It relies on the ISO standards for life-cycle assessments (ISO 14040 and ISO 14044, see below, this chapter) to specify the requirements for GHG emission analysis for products over their entire lifecycle. A revised version was published in 2011 (BSI, 2011).

Of course, this methodology, which is based on the LCA concept, suffers from similar limits to those shown in the section on LCAs. The results obtained from applying PAS 2050 present a high level of uncertainty (rarely below 10%, and sometimes above 30%), particularly due to the subjectivity of the rules for allocation and use of secondary data.

The PAS 2050 specification stipulates that primary activity data should always be used if it is available or could be obtained at a reasonable cost (which is clearly never the case). Where other figures are used, they should be justified.

If a process that is responsible for a portion of emissions produces more than one useful sub-product, the emissions should be allocated (see "Allocation of emissions among co-products" above). The specification offers recommendations on this subject. If possible, the process should be divided into different sub-processes (for example isolating the machines required for processing). If this is not possible, the boundaries of the system should be extended to include the impact of the other co-products (system extension). If allocation is necessary, it is recommended that GHG emissions are allocated in proportion to the economic value of the co-products. With this method, if use of carbon from fossil fuels is avoided by co-generation, this should be taken into account within the expanded system, provided the energy is fed back into a national grid or on a grand scale. Attribution between electricity and heat can be made using a recommended ratio.

The PAS 2050 specification states that if a product requires re-emission of biogenic carbon into the atmosphere during the period of evaluation, set at 100 years as per the IPCC recommendations, this stored carbon should take into account pro rata for the storage period in relation to a century.

The following elements must also be excluded from the boundaries:

- sources of negligible emissions (<1% of the total impact) up to 5% of the total;

- human contributions to the process (e.g., maintenance).

The Carbon Trust has also published a code of good practices for communicating information on carbon footprints and a guide for implementing the standard.

The GHG Protocol Product Life-cycle Accounting and Reporting Standard

This standard was originally inspired by the specification PAS 2050. The PAS 2050 specification was revised in 2011 to take into account lessons learned while developing the GHG protocol for products, and to ensure that these two standards covering the same subject were coordinated. The two standards are consistent, but each retains its own specific characteristics. For example, the PAS 2050 specification provides recommendations on how to record data and the GHG protocol for products offers recommendations on com-municating products' emissions to the public (WRI and WBSCD, 2011b). Given that the two standards are compatible and comple-mentary, applying the principles of both standards when calculat-ing the carbon footprint for a product is recommended.

Towards international standardization?

The ISO (International Organization for Standardization) is an international standardization body made up of representatives of national standardization organizations from 162 countries. The most influential members of the ISO include the German Standardisation Institute (DIN), the American National Standards Institute (ANSI), the Association française de normalisation (AFNOR – French Standardization Association), the Japanese Industrial Standards Committee (JISC) and the British Standards Institution (BSI).

The ISO produces voluntary standards for sectors and stakeholders who have an explicit need. ISO standards are produced by technical committees of experts from the industrial, technical and economic sectors. These experts may include representatives of government agencies, consumers associations, NGOs and universities. The ISO standards are voluntary, and they are the result of consensus among international experts. The principle of consensus requires that all objections are examined in depth and a response is given. This is a fundamental element of the procedure for producing standards.

The ISO has been a leading source of standards, technical reports and technical specifications relating to inventory and control of GHG emissions. Standards and technical specifications and reports ISO 14064, ISO 14065, ISO 14066, ISO/TS 14067 and ISO/TR 14069 provide a recognized framework for the international plan for measuring GHG emissions, verifying emission declarations and providing accreditation for bodies active in this sector. The ISO has played a vital role in harmonizing carbon accounting, and the ISO standards have made a contribution to the development of legislation at a national and regional level (e.g., the European law on monitoring of GHG emissions) and even at the international level (providing recommendations used within the framework of emissions monitoring for the UNFCCC and the Kyoto Protocol). Before it began focusing more attention on carbon in the 2000s, the ISO first published standards on LCAs during the 1990s. The following pages provide an introduction to the main ISO standards for carbon accounting.

ISO 14025: standards on environmental labels and declarations

ISO 14025:2006 establishes the principles and specifies the procedures for developing environmental declarations. It specifically establishes the use of the ISO 14040 series of standards in the development of environmental declarations (ISO, 2006f). Environmental declarations as described in ISO 14025:2006 are primarily intended for use in business-to-business communication, but their use in business-to-consumer communication under certain conditions is not precluded.

The standards ISO 14040 and 14044: standards on life-cycle analysis

The standard ISO 14040 describes the principles and framework for LCAs, as well as good practices for carrying out this type of study (ISO, 2006a). First published in 1997, a revised version came out in 2006. The standard ISO 14044 specifies requirements and provides guidelines for definition of objectives and the area of study, the inventory phase, the impact analysis phase, the interpretation phase, communication and critical review of the life-cycle analysis (ISO, 2006b).

According to the ISO standards on LCAs, the choice of impact categories cannot cause double accounting or mask significant environmental impacts. Although these standards provide certain recommendations on how to determine the allocation rules, the users of the standards are still allowed a lot of freedom of interpretation.

The standard ISO 14044 stipulates that if the results of the LCA are based on a process of weighting (single score, normalization, monetization etc.), the results of this process must be presented together with the unprocessed results, using appropriate physical units that are in common use (CO_2 equivalent for the greenhouse effect, H+ equivalent for acidification etc.).

Finally, according to the ISO standards, if an LCA is published it must be reviewed by external auditors.

ISO 14064: the standard for GHG emissions inventories

Published in 2006, the standard ISO 14064 comprises three parts, each representing a different technical approach. The first part provides indications of the elements required to establish an inventory of verifiable GHG emissions. It provides a framework for design, adjustment, management and declaration of GHG inventories by organizations (ISO, 2006c). The standard starts by presenting the general principles of carbon accounting. It was drawn up after the Greenhouse Gas Protocol, so its general principals are very similar: relevance; completeness; consistency; accuracy and transparency (see Box 2.1, The general principles of carbon accounting). The standard then deals with setting the boundaries, which include organizational boundaries and operational boundaries.

Organizational boundaries relate to entities in the organization that should be included in the inventory. There are two possible approaches for setting these boundaries: one can consolidate emissions on the basis of control or on the basis of share ownership. With the control approach, the organization carries out the GHG inventory for all the entities it controls (financially or operationally).

With the ownership-based approach, emissions are reported proportionally on the basis of share ownership.

As we have seen in the section on the concept of boundaries, operational boundaries relate to activities covered by the inventory. Direct GHG emissions (scope 1) should always be included in the inventory. Indirect emissions from energy production are also always included (scope 2). Other indirect emissions, such as emissions for employees' travel in vehicles not controlled by the company (scope 3) can be optionally included.

This standard includes requirements and guidelines on inventory quality management, writing reports, internal audits and the verification responsibilities of the company.

The second part of the standard, ISO 14064-2, deals with quantification and reporting of emissions linked to project activities (ISO, 2006d). In this sense it was driven by the rapid development of compensation projects following the Marrakesh Accords. This methodology has mainly been used for developing offsetting projects, particularly in the framework of the Clean Development Mechanism (CDM, see Chapter 6).

ISO 14064-3 lists the principles and requirements for verification of GHG inventories and validation or verification of GHG projects (ISO, 2006e). It can be applied to GHG quantification at the level of the entity or project, and specifies the requirements and recommendations for people who want to carry out GHG validations and verifications. This verification standard is applicable to independent or internal auditors. Unlike the first part, which was largely inspired by the Greenhouse Gas Protocol, the third part is an original document conceived by ISO and based on best practices in carbon accounting and environmental auditing, as well as early verification experiences in the CDM and the first carbon markets. The basic principles for verification include independence, respect for ethics, presenting a true picture, and professional conduct during the mission. Verification also involves defining a threshold of materiality so one can judge if the errors, omissions or inconsistencies in the declarations or data are significant or not.

ISO/TS 14067: a technical specification for product carbon footprints

ISO/TS 14067 is a new international technical specification. Released in 2013, it covers the carbon footprints of products and how those footprints are communicated, including labelling. It is largely inspired by PAS 2050 and the GHG Protocol for products (ISO, 2013). It defines the carbon footprint of a product as the sum of GHG and removals in a product system, expressed as CO_2e and based on a life-cycle assessment that uses the single impact category of climate change. Offsetting, for instance, through an investment outside the relevant product system (e.g., investment in renewable energy technologies, energy efficiency measures, afforestation/reforestation) is not allowed when quantifying a product's carbon footprint. The communication of offsetting is explicitly outside the scope of this standard (the PAS 2060 is a British specification that addresses this topic).

ISO/TR 14069: a technical report detailing scope 3

ISO/TR 14069 is a technical report that supports the application of the international standard ISO 14064-1 for quantification and declaration of organizational GHG inventories, particularly regarding scope 3 emissions (ISO, 2013).

Despite the development of this technical report, accounting of indirect emissions is still the subject of significant debate. As it is difficult to collect data from suppliers, economic actors are concerned that the accounting and publishing efforts involved with this scope lack relevance. Some companies are also worried that they may be held responsible for the emissions of their suppliers, although they do not always have reliable information on these emissions and they may be unable to influence them.

Towards global carbon disclosure for products?

(Some) pressure from retailers

In the context of increased communication about the environmental performance of products, sometimes with a hint of *greenwashing*, carbon accounting is going beyond the circles of experts and people who read company annual reports, and emissions are now being displayed on product packaging. This trend started in the United Kingdom, when the Tesco supermarket chain launched a programme to calculate the carbon footprint of consumer packaged goods. The programme aimed to eventually cover 70,000 products and was inaugurated in 2008 when carbon footprints were calculated and published for 30 products from five families. This preliminary work was carried out in collaboration with the Carbon Trust, and it provided an opportunity to test the PAS 2050 specification. However, despite significant investment, particularly in communications, the British retailer dropped this initiative in 2012. The programme was found to be too costly, particularly as other retailers did not follow the initiative. Surveys also showed that consumers found the label too complicated, undermining the competitive advantage that this type of disclosure might have offered (see Figure 4.2).

In France the methodology used by the retailer Casino to determine the carbon index for products is also inspired by LCAs, but it excludes the phases of consumption and extended use. This programme is undoubtedly the most ambitious in terms of communication, but although it is inspired by the ISO standards the results are still largely based on secondary data and they have not been subjected to third-party review. This limits the relevance of the results (see Figure 4.2 for an illustration). However, an advantage of this approach is that it can highlight the products with the most significant impacts (for example meat has a more significant climatic impact than potatoes), although it does not enable any comparison between products from the same category.

FIGURE 4.2 Carbon disclosure among large retailers (voluntary initiatives in the United Kingdom)

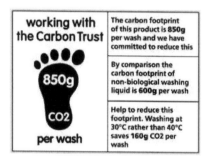

Towards a standard for environmental disclosure for products?

France is probably the country that has invested the most in environmental disclosure of products. BP X30-323 is a public reference document that was created by AFNOR at the time of the Grenelle environment forum. It describes general principles for environmental disclosure and the methodological framework for estimating the environmental impact of consumer packaged goods. The aim is that this information should become a criterion for consumer choices. According to this standard, the information should be visible to the consumer, easy to read, comparable between products and identical everywhere, in order to make it relevant. Information relating to calculation of indicators should be transparent to ensure the credibility of the figures and comparability between products.

BP X30-323 came out in June 2011. Following its publication, an experimental project started with the participation of just over 160 companies in France. There are three key differences between this standard and others that are being developed (mentioned in this chapter). Firstly, this methodology is part of a binding legal framework (the Grenelle Act) whereas the other approaches and standards considered here are voluntary. Secondly, the aim in this case is not just to carry out an LCA or calculate a carbon footprint, but also to communicate the result in order to influence consumer choice. Finally, the disclosure does not focus solely on the carbon footprint, as it has several criteria. Other relevant indicators are defined for each category of products. For example, toxicity may be relevant for a detergent, or exhaustion of fish stocks in the case of fish.

According to a statement from the Centre d'analyse stratégique (CAS – Centre for Strategic Analysis) issued at the start of 2013, France should pursue its strategy of supporting environmental disclosure, in particular by announcing obligatory labelling on consumer goods within three to five years. However in 2016 no mandatory laws were passed. The CAS statement started by underlining the need to continue with concerted experiments to raise the profile and credibility of the label, and make it easy to identify and simple to interpret.

Such support from the authorities for an environmental disclosure initiative is unique, and it raises the question of their motivation. Are they trying to boost the weak carbon balance of products manufactured in France, taking advantage of the significance of nuclear energy? Are they trying to introduce a market barrier for foreign producers or those who do not have the resources to calculate the impact of their products? Is product disclosure first and foremost an effective way to encourage the public to adopt more sustainable consumer choices? The last hypothesis seems doubtful in the light of studies on the experiences of Tesco in the United Kingdom and the marketing strategies of retailers. These companies often favour the most lucrative products (such as red meat or exotic fruits) over low-carbon products that generate low revenue (for example local, in-season vegetables).

Disparate country-level approaches are an issue for large corporations, free-trade areas and international organizations such as the EU or WTO because of the market distortions it can create. Imagine, a given company wishing to market its product as a green product in UK, France, Italy and Switzerland. Would it need to apply different schemes in order to compete based on environmental performance in the different national markets? In France, it would need to carry out an environmental assessment in line with the French method (BP X30-323), in the UK, it would need to apply the PAS 2050 or the WRI GHG Protocol, in Switzerland, it would need to apply the Swiss approach (currently under development), and in Italy, it would need to join the governmentally recognized carbon-footprint scheme, and carry out yet another analysis. The same company would also need to develop an Environmental Product Declaration (EPD) according to ISO14025:2006 for the Swedish market. They may then need to undertake multiple EPDs as there are at least six competing EPD systems around the world with their own specificities.

5

Monetary inventories: what is the cost of a tonne of CO$_2$?

Since the early 1990s when people became more aware of anthropogenic influence on climate change, economists have tried to attribute a monetary value to GHG emissions. In the 2000s, various countries including France and the United Kingdom funded work to estimate or set the value of carbon. Supporters of the monetization approach believe that this value is necessary in order to take this environmental externality into account in investment decisions and to set or harmonize climate policy and measures (taxes, grants etc.). As will be shown, this approach can still be criticized for its level of uncertainty and strong assumptions. The main criticism relates to the utilitarian and anthropocentric roots of techniques for monetizing environmental impacts, which are based on the premise that we have no moral obligation to nature (Milanesi, 2010). Despite this criticism, it is a vital element for implementation of the polluter-pays principle, one of the key principles in environmental law (see Chapter 6).

There are basically two types of analysis to determine the cost of a tonne of CO_2. The first approach involves analysing the costs and benefits of a climatic action in order to determine the economically optimal point where the benefits of an action are equal to the costs. This was the approach chosen by the authors of the Stern Report in the United Kingdom (see Box 5.1, The Stern Report on the economics of climate change). The second approach involves minimizing the cost of climatic action by setting a reduction goal. This is a cost-effectiveness analysis perspective, along the lines of the Quinet Report on the shadow price of carbon (2008).

The two approaches are complementary. The cost–benefit approach attempts to define the optimum level of greenhouse gas concentration in the atmosphere, whereas the cost-effectiveness approach attempts to link a given initiative to a consistent carbon value with the aim of reducing emissions. In addition to these two theoretical approaches, the cost of carbon can be determined by considering the values revealed on carbon markets or the values imposed through an environmental tax (see Chapter 6). In this case the cost of carbon depends on expectations of future climate policy, particularly the way in which the EU will implement its commitments to reduce emissions in different sectors.

Cost–benefit analysis of climate action

It is a vital part of economists' work to put the costs and benefits of everything into perspective. According to economic theory, cost–benefit analysis should guide decision-makers' choices. This approach lists the positive and negative effects of an action or a project on well-being, and then assesses them in monetary terms (Blanchard and Criqui, 2000). From this perspective, the aim of environmental remediation is to maximize the net benefit of the actions carried out. In the context of the fight against climate change, the benefit is equivalent to the cost of damages avoided thanks to a reduction in GHG emissions (Nordhaus, 1991). A tonne of CO_2 emitted today is associated with the value of the future damage to be assessed. Optimum environmental remediation is to be found at the point where the marginal cost of emissions reduction (MAC), that is, the additional cost incurred for the last unit of CO_2-equivalent reduction, becomes equal to the marginal benefit associated with this reduction, that is, the additional benefit caused by the last unit of CO_2-equivalent reduced. This optimal environmental remediation point is the Social Cost of Carbon (SCC, see Figure 5.1). The SCC enables the determination of benefits of a given climate policy from a cost–benefit analysis perspective. Figure 5.1 shows that the marginal benefits of reduction decrease as reduction objectives increase. At the same time the marginal abatement cost increases as the reduction objectives become more ambitious.

FIGURE 5.1 The social cost of carbon

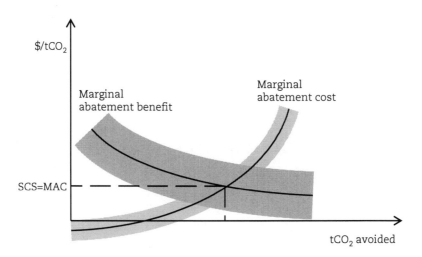

In practice, it is complicated to apply this theoretical approach as it is difficult to put a figure on the damage caused by GHG emissions at a global level. This process raises methodological, ethical and philosophical difficulties.

Monetary assessment of climate change is carried out using integrated impact-assessment models (Hope, 2005). First, demographic, economic and technological trends are analysed to determine future emission levels (with a time horizon set *ex ante*). These levels then make it possible to determine the concentration of GHGs and to deduce their radiative and climatic impact. The projected impact is then analysed at a regional level to simulate the temperatures and rainfall region by region, and estimate the consequences for cultures and natural ecosystems. The final stage involves setting a value for these impacts, both market-based (improvement of agricultural yield in some regions, degradation in others etc.) and non-market-based (loss of human life due to natural disasters or droughts, deterioration of ecosystems etc.). This type of analysis, which is generally presented in a linear fashion, cannot provide entirely satisfactory answers, as the socio-economic impacts estimated at the end of the process influence the parameters set at the start of the process. If the projection within the time horizon studied indicates a significant demographic and economic impact, this has a considerable influence on the predictions for emissions for a longer time horizon.

FIGURE 5.2   **Integrated impact assessment model**

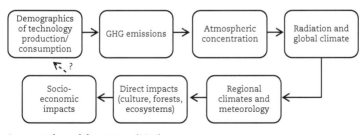

Source: Adapted from Hope (2005).

One of the most significant difficulties stems from monetary assessments of biodiversity and ecosystems. Inventories of potential physical damage linked to climate change are subject to significant scientific uncertainty, and many of these impacts are non-market-based (IPCC, 2014). No market can really enable one to assess the value of the aspects considered. Also, this approach implicitly relies on the premise that natural capital and economic capital are interchangeable. Ethical questions arise when determining the statistical value of a human life in an international approach, or carrying out a monetary assessment of impacts on biodiversity. These questions are complicated by the fact that it is difficult to compensate for these costs with the advantages that certain areas could gain from temperature rises. The geographical and temporal division of costs and damages is only rarely included with the results of cost–benefit analysis.

The final methodological obstacle for this type of economic analysis concerns the choice of discount rate. This rate plays a fundamental role in disclosure of damages with a long time horizon. Due to interest rates and inflation, a Euro in 50 years' time would have a lower net value than a Euro today. For example, 1 million EUR of damage in 50 years discounted at a rate of 5% represents only 87,200 EUR today. This is one of the pitfalls of standard economic analysis: its short-sightedness with regard to long-term political choices.

Recent impact assessments establish the cost of annual damage caused by doubling the concentration of CO_2 in the atmosphere as a few percentage points of GDP. However, the cost would be proportionally a lot higher for developing countries (up to 10%) than for industrialized countries (around 1% to 3%). The impact could also be more marked if GDP does not grow in line with scenarios established before the economic crisis of late 2008. The values estimated by Stern (2006) in case of inaction are even higher, largely because he chose a lower reduction rate (see Box 5.1).

Despite these limits, cost–benefit analysis is still useful, as it provides an approximate picture of the issues and the initiatives that must be implemented, to an order of magnitude. These values are particularly interesting for insurers, who want to internalize a certain number of climate change effects that could affect their clients (natural disasters, pandemics etc.).

Nevertheless, cost–benefit analysis should be used with care in order to ensure that economic incentives are set at appropriate levels (Quinet, 2008). The value of carbon does indeed go down as emission-reduction goals rise, as less significant damage will be caused if there are ambitious reduction goals. For example, in the Stern report, the social cost of carbon is estimated at \$85 per tonne of CO_2 in the *business as usual* scenario. The cost would be only \$25 with a goal of stabilising at 450 ppm CO_2 equivalent.

Box 5.1   The Stern report on the economics of climate change

Published on 30 October 2006, this study headed by Lord Nicholas Stern on "The economics of climate change" (Stern, 2006) is the first substantial report on global warming carried out by a team of economists.

Despite the hybrid nature of the document commissioned by the British government, which mixes scientific expertise and political argument, it gains legitimacy from the significant review of academic literature. The document is more than 600 pages long.

The report includes a cost–benefit analysis of climatic action and it attempts to set an optimal level for global restriction of emissions in order to balance the marginal cost of reducing a tonne of CO_2 and the discounted amount of marginal future damage caused by a tonne of CO_2 emitted today. As damage may occur over a very long time horizon, the discount rate selected is a strong assumption.

Stern chose a very low discount rate (1.4%) on ethical grounds (protecting the climate for future generations), and came to the conclusion that the benefits of policies and measures are much greater than the costs. In his conclusion, Stern estimates that if no action is taken, the damage caused by climate change will be equivalent to a loss of at least 5% of global GDP per annum. If one takes into account a greater range of risks and consequences, the estimate could rise to 20% of GDP or more. On the other hand, the cost of action to reduce GHG emissions in order to avoid the worst consequences of climate change could be limited to about 1% of global GDP.

We should bear in mind that impact assessment is complicated by uncertainties regarding the consequences of climate change and the absence of a homogeneous and ethically acceptable method to express the impacts of climate change in monetary terms, particularly for a long time horizon, which is greatly frowned upon with respect to economic analysis. Also, it is not enough to propose a single figure, as it is necessary to know how the cost is distributed over time, geographically and among different social classes.

Cost-effectiveness analysis of climate action

With cost-effectiveness approaches, the value of carbon is determined by the cost of the last tonne of CO_2 reduced to achieve a set or prior reduction objective (i.e., the marginal cost of reduction).

This value depends on the level of environmental remediation for the objective. In other words, the value of carbon in cost-efficiency analysis ultimately corresponds to the cost of the last process used to achieve the emissions objective (Quinet, 2008).

The MAC is generally presented in the form of a curve showing how it increases for each unit of emission reduction (see Figure 5.3). Indeed, it is most economically and socially effective to deploy the least costly reduction measures first. This curve can be deduced from consultations with experts, who assess the cost and emission reduction potential of given technologies. The consultancy firm McKinsey has popularized the use of these curves in many countries and industrial sectors.

An interesting characteristic of this curve is that there is an initial section of negative costs. From a technical perspective, this means that there is a potential for risk-free energy savings, which could actually be lucrative without costing CO_2 (e.g., standby mode on electronic equipment or building insulation). From a macroeconomic perspective, one could say there is a double dividend (economic and environmental) for these negative costs. These lost opportunities show that markets are far from perfect and that the actors do not always maximize their profits. Risk-free opportunities, such as building insulation, have not been taken up for various reasons: user habits (users prefer products that they are familiar with and don't trust new products), inflexibility of technology (e.g., the path dependence of systems based on fossil fuels), imperfect information (Maréchal, 2007). In addition, some economists have questioned the existence of negative reduction costs. Indeed, this anomaly could also mean that certain costs have been omitted from the analysis, such as the cost of accessing information, the difficulty of gaining access to credit, transaction costs (e.g., in the relationship between renters and lessors). (Quinet, 2008).

FIGURE 5.3   Stylized example of marginal costs of GHG
emission reductions

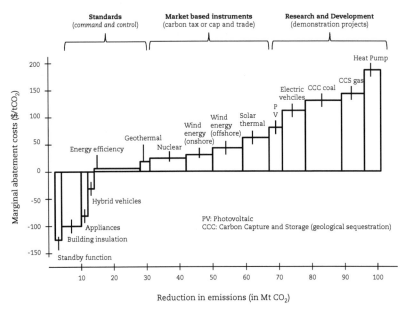

Source: Kesicki and Strachan (2011).

It is particularly important to determine the curves in different sectors of activity, and indeed to be aware of them, in the framework of climate policy that relies on market-based instruments. Manufacturers who know their reduction costs could adopt strategies to reduce overall reduction costs, and only invest in the most lucrative reduction measures. However, in Europe it seems that many manufacturers who are subject to quotas do not know their MAC, which in turn limits their arbitration options and the economic efficiency of the system (Engels, 2009).

In addition to analysis of reduction costs in different sectors, one can use economic models that bring together the entire economic system of a country in order to calculate a carbon value that is consistent with the emissions level that should not be exceeded. These models require assumptions about economic growth, the emergence of new technologies, the response of economic agents to a price rise for raw materials etc. The difficulty here lies in behaviour modelling in response to prices, the learning curve for technologies and assumptions about future innovations (Quinet, 2008).

The EU has currently set itself the objective of limiting temperature rises to 2°C. To have at least a 50% chance of achieving this objective, the concentration of GHGs must not exceed 450 ppm of CO_2-equivalent (IPCC, 2007). In this context, it is no longer necessary to refer to an analysis of the greenhouse effect damage. We assume that these elements, along with other scientific knowledge, have been taken into account in the course of defining political objectives. For this reason, the commission chaired by Alain Quinet in France chose to follow this approach when defining the shadow price of carbon.

Towards a shadow price of carbon?

Values attributed to CO_2 that is emitted or avoided are comparable to values attributed to time that is saved or lost. They are referred to as shadow prices, partly because they are not provided directly by a market, and partly because the market prices provide an imperfect reflection of real social costs (Baumstark, 2007). Shadow prices are set by the state on the basis of public expectations, so they can be understood. In general, they are the result of a compromise made by a planning authority on the basis of a dialogue between economists, economic and social partners, and representatives of NGOs.

If public authorities do not know the cost of CO_2 when they make decisions, they risk supporting policies that exacerbate climate change. In the United Kingdom, the Department of Energy and Climate Change (DECC) set the shadow price of carbon at £60 in sectors not subject to CO_2 quotas (DECC, 2009). This value will rise progressively (reaching £200 in 2050). However the DECC may have to revise these figures as it estimates the level of uncertainty to be 50%. For sectors that are covered under the EU, ETS-suggested prices are much lower, at about £6 in 2020 (DECC, 2015).

There is no official shadow price in the United States or Canada, but a growing number of companies (97 in 2015) are assigning an internal price to their carbon emissions. These include consumer brands such as Colgate-Palmolive and Campbell's Soup, global industrials such as General Motors, and financial giants such as TD Bank (CDP, 2015). Globally, 435 companies reported using an internal price on carbon in 2015, up from only 150 in 2014 (CDP, 2015). A variety of drivers are cited, including incentivized investments in clean energy and emissions reductions, and the need to mitigate risk from future regulation and global carbon pricing frameworks. Global companies have voluntarily enacted carbon pricing as a way to address mandatory carbon pricing enacted via regulatory regimes in other regions.

In France, the Quinet Commission report analysed the progression of CO_2 values required to achieve European political objectives, that is, a 20% reduction in 2020 and an 80% to 95% reduction in 2050 compared to 1990 (Quinet, 2008). In this case, the shadow price changes from 32 EUR per tonne of CO_2 in 2010 to 200 EUR in 2050 (a price range of 150 – 350 EUR).

Although this reference value is imprecise and limited (see Table 5.1), it does make it possible to put a value on the greenhouse effect when assessing the profitability of investment projects and the effectiveness of public policy. The shadow price also provides a signal that can guide companies' R&D and investment strategies as more and more corporations are using an internal price of carbon to manage their climate risk exposure. Although the shadow price does not directly affect the financial result of private decisions, it provides information on the ambitions (and constraints) that have been set for the medium and long term. This sends a signal that developing new low-carbon technologies may well be profitable (Quinet, 2008).

TABLE 5.1   **A highly unpredictable carbon price depending on the working assumptions**

Social cost of carbon	Uncertain value (between 1 and 1,000 EUR, according to currently published estimates). $85 in 2000 according to the Stern Report in a business-as-usual scenario (Stern, 2006).
Marginal abatement cost	Uncertain value, sometimes negative, sometimes hundreds of dollars, depending on the type of technology considered and how ambitious the reduction objectives are.
Shadow price of carbon	In France, 32 EUR per tCO_2 in 2010, rising to 200 EUR in 2050 (price range 150 – 350 EUR). In the United Kingdom the value changes from £64 in 2020 to £212 in 2050.

6

Internalizing physico-chemical inventories in economic and political decisions

Various complementary tools

Standard economic theory considers GHG emissions to be a negative externality, that is, a cost that is not reflected in the market price but that affects economic third parties who do not have a legal right to compensation. The market is unable to attribute a market price to these undesirable external effects spontaneously, and they are a source of inefficiency: the equilibrium quantity does not maximize the surplus for consumers and producers (total surplus). When there is an externality, the social cost of production is higher than the cost borne by the producers.

By the polluter-pays principle, this market failure requires the state to intervene using legal, fiscal or economic instruments. The polluter should take responsibility for the "expenses of carrying out the (...) measures decided by public authorities to ensure that the environment is in an acceptable state" (OECD, 1972).

Traditionally, the most commonly used instruments have been regulations or binding standards ("command and control" instruments). Regulations to limit externalities include banning emissions of certain substances (such as substances that damage the ozone layer within the framework of the Montreal Protocol), requiring certain levels of performance for industrial facilities, setting noise standards and banning electrical devices that are too inefficient.

Another solution is to implement fiscal incentives, that is, introduce taxes or subsidies. These are known as Pigovian taxes, named after the economist who proposed the first taxes to internalize externalities (Pigou, 1920). Duties levied on alcohol, petrol or tobacco are specific examples of these taxes. Subsidies can be used to internalize positive externalities.

More recently economists have proposed a third way of internalizing environmental externalities: defining property rights for the externality and setting up a market in which to trade these rights. Thomas Crocker and John Dales were the first to recommend implementation of this theory in the late 1960s, proposing that a quota-trading system should be set up to combat environmental pollution (Crocker, 1966; Dales, 1968). Crocker proposed implementing a trading system to limit emissions of atmospheric pollutants from fertilizer factories in Florida, while Dales was concerned with farmers polluting the Great Lakes between Canada and the United States. These ideas were not implemented until 25 years later, when markets were set up to trade sulphur dioxide–emission rights in the United States in the framework of the programme to combat acid rain (Joskow and Schmalensee, 1998).

The aim of emission-rights markets is to achieve a given reduction in emissions for a minimal cost. This mechanism involves determining *ex ante* an overall maximum quantity of pollutants that must not be exceeded (for example GHG emissions) and distributing this allocation among different economic actors. With this system, any agents who decide not to use all their emission rights can sell unused rights to another agent. This agent is then authorized to emit more pollutants, in line with the quantity of emission rights purchased. This type of trade is beneficial for both parties involved, as the marginal abatement cost for one agent is lower than for the other. It is indeed less costly for the purchaser to buy rights than to reduce emissions, and it is profitable for the other agent to reduce more emissions in order to be able to sell rights. Emission-rights trading is a form of compensation, where the value is determined by the cost of reduction avoided by purchasing rights and the reduction cost actually incurred by the vendor of the right. Making the rights negotiable enables profits to be made and thus reduces the total cost of reducing emissions.

An emission-rights market offers more certainty that environmental objectives will be achieved than a Pigovian tax. However, this increased certainty brings with it an uncertainty about emission-rights prices, which could in theory be very high, thereby encouraging competition among companies. The uncertainty could equally discourage investors due to the implicit risk. As we will see, the choice between carbon markets and taxes is essentially a political one.

Finally, the end of the chapter will examine voluntary initiatives to internalize the cost of carbon by analysing voluntary carbon offsetting.

Carbon taxes

Environmental taxes reduce emissions by modifying behaviour and encouraging development of new technologies (Gerlagh and Lise, 2005). Carbon taxes are taxes assessed on the basis of the carbon content of energy sources (Elbeze and de Perthuis, 2011) and rates are expressed as a monetary unit per tonne of CO_2. These taxes aim to change the relative prices of goods or energy sources, on the basis of their CO_2 content, in order to steer decisions made by actors towards low carbon production and consumption. This type of tax may be primarily motivated by a desire to regulate diffuse emissions, however other factors explain its appeal to many governments. For example, it is a new source of resources for the state, particularly as company tax rates have dropped. Supporters of carbon taxes argue that tax structures should be modified to reduce taxes deemed to cause distortions, such as taxes on earned income (Bosquet, 2000). For fossil fuels, carbon taxes are also a way of intercepting rents on oil and other fuels (Elbeze and de Perthuis, 2011). At this point, it is important to be clear about the fundamental difference between carbon taxes and energy taxes. Whereas carbon taxes have a direct effect on CO_2 emissions, energy taxes aim to rationalize energy use whatever the carbon content. A legislator can decide to tax these two areas in a complementary way.

An increasing number of European countries and Canadian provinces have imposed a tax on CO_2 emissions. Finland was the first country to introduce a carbon tax in 1990. Sweden, Denmark, the Netherlands and the United Kingdom followed, with reforms that moved the fiscal burden from labour to CO_2 emissions and energy consumption. These environmental taxes have increased progressively in Europe, now accounting for 25 billion EUR of income, which has made it possible to reduce labour taxation (Ekins and Speck, 2011). These countries were joined more recently by Switzerland (2008) and Ireland (2010) (Elbeze and de Perthuis, 2011). During this time, the tax rate has increased progressively but significantly. In Sweden, the tax rate for households has risen from around 25 EUR in the early 1990s to more than 100 EUR per tonne of CO_2 in 2011 (Elbeze and de Perthuis, 2011). It was possible to impose a high level of taxation in the Scandinavian countries as income tax and social contributions were reduced at the same time (*green tax shift*). We note that a carbon tax of 100 EUR per tonne is still low compared to the level of energy taxation already applicable to fuels; the carbon tax increases the price of a litre of diesel by only 26 cents. In France, the plan to introduce a carbon tax known as the "contribution climat-énergie" (climate-energy contribution) within the framework of the Grenelle environment forum was finally abandoned in March 2010.

Traditionally, one of the difficulties in implementing environmental taxes is the huge range of natural resources to protect. The many carbon tax initiatives were made possible by defining a common measure, the tonne of CO_2 (see Chapter 2). This unit of measurement is clearly a huge asset when developing economic instruments, especially as there are multiple emissions sources with very different reduction costs. However, the economic benefits of carbon taxation compared to a regulation approach are so much higher that there are significant differences between the cost functions of the different actors who need to reduce their emissions (Elbeze and de Perthuis, 2011).

The idea of a carbon tax applicable to all EU member states was first proposed by the Commission in 1992. The objective was to promote a move from purely regulatory tools (*command and control*) towards economic tools thought to be more efficient. However, this proposal, and the amended version two years later, was rejected by member states as they were reluctant to grant fiscal powers at a supranational level or they feared for their industrial development. It is worth noting that in community law, unanimity is a requirement for implementation of fiscal powers. Following these failures, the Commission favoured implementation of a carbon market to limit emissions from industry. However, more than half of European emissions are not covered by this system, which explains why many member states have pursued or adopted their own carbon tax system.

In 1997, the EU tried another approach aimed at harmonizing energy taxation systems (Speck, 2008). The proposal emerged in 2003 with the adoption of Directive 2003/96 restructuring the community tax framework for energy and electricity products (EU, 2003a). Nevertheless, the text of this directive was highly softened in comparison to the Commission's initial aims. Specifically, this text stipulates that member states should have a minimum level of taxation for certain energy products, notably natural gas, coal and electricity. The directive also distinguishes between heating fuel and motor fuel. Thus, although the minimum charge on unleaded petrol (motor fuel) is set at 359 EUR per 1,000 litres (more than 120 EUR per tonne of CO_2), the charge for fuel oil (heating fuel) is just 15 EUR per 1,000 litres (barely 5 EUR per tonne of CO_2). The minimum tax threshold for natural gas is twice as high for domestic use as for professional use. In 2012–13, this text was revised in order to increase the thresholds and further penalize the biggest CO_2 emitters, particularly coal.

Taxes or a carbon market?

Economists have shown that in a world with perfect information, where the state knows the emission reduction costs for companies, it can choose to authorize a quantity of emissions (quotas) or to set its price (tax), leading to the same reduction in emissions (Weitzman, 1974). In both cases, the price of carbon would theoretically make it possible to implement the cheapest technologies in order to reduce CO_2 emissions to a certain level. However, in practice, information is not perfect, and the regulator does not know the marginal abatement costs. If the regulator has to choose the tax rate or the quantity of quotas to grant without knowing the marginal abatement costs, they will not be able to ascertain the optimum level of reduction that would maximize social benefits. The major disadvantage of putting a fixed quantity of emission quotas into circulation is that there is no price guarantee (and therefore no guarantee of the costs for companies). *A contrario*, although a tax enables the price to be known, the unknown factor is the quantity of CO_2 that will be emitted.

Economic theory indicates that it is better to use a tax if the marginal abatement cost increases more quickly for enterprises than the marginal environmental damage (i.e., if the gradient of the cost curve is steeper than the gradient of the damage in Figure 5.1). This is the case for pollution that accumulates to form a stock, such as CO_2 in the atmosphere. Each additional unit of pollution will have a more or less identical marginal impact on the environment, while the marginal cost of reduction could increase much more significantly, in cases where significant technological jumps are necessary to further reduce emissions. *A contrario*, it is worth using cap-and-trade systems if the need for certainty about the quantity of pollution emitted is more important than the cost for the economic actors, that is, if there is an ecological threshold that must not be exceeded (Weitzman, 1974). Therefore, taking into account the risk of tipping points with irreversible reactions due to human influence on climate may mean it is better to implement a quota system. Clearly we do not know the precise level of CO_2 for these thresholds, but we can consider the limit of 2°C/450 ppm to be an approximate threshold not to be exceeded, in order to avoid such tipping points. It is also possible to combine taxes and carbon markets. There could be a tradable quota system where the price of quotas was capped by a tax (Cournede and Gastaldo, 2002).

Constituent elements of a carbon market

Legislators setting up an emission-rights market start by defining the desired emission level *ex ante*. In this market, demand is created by emission caps, and any quantity of emissions must be covered by an equivalent emission quota. Demand for quotas or credits would therefore depend on the severity of the cap in comparison to the actual emission level for the agents involved.

Then the legislators have to define and allocate the emission rights representing the emission objective. These emission rights, known as quotas, correspond to the supply of emission rights on the market. As these rights serve to cover the actual emissions of the states or industries concerned, when an agent reduces emissions, this increases the supply of rights available for other agents. For the defined period, the emission rights should cover actual emissions and be reconciled with the binding objective.

There are different methods for allocating quotas: free on the basis of historical emissions, free on the basis of energy efficiency benchmarks, or a price may have to be paid through a system of bidding. The first two methods are supported by industry, as they do not involve transfer of money from the private sector to the public sector. *A contrario*, environmentalists reject these free allocations of emission rights, which they see as a "right to pollute", and they condemn the privatization of a common resource (the composition of the atmosphere). An approach based on historical emissions rewards big historical emitters, while allocation on the basis of benchmarks favours those who have already significantly reduced their emissions in the past or new actors who can invest directly in the best technologies available. In practice, the political context often encourages legislators to allocate quotas using a combination of the three approaches.

When setting up a cap-and-trade system, legislators often implement **project-based mechanisms**. Thanks to such mechanisms, an investor in low-carbon projects can generate additional emission credits from sectors or regions not covered by the cap system. The emissions savings from the investment give the investor the right to emission credits when they have been checked by an external auditor. These credits authorize additional emissions by the agent that owns them, and the mechanisms make it possible to reduce observance costs by increasing the supply of emission rights.

Monitoring and reporting of emissions are vital elements. Indeed, one can only know if the environmental target has been reached after calculating actual emissions. It is vital for the credibility of the system to define clear and standardized rules for the emission-calculation method. As emission rights are fungible, it is important that the calculation methods are reliable, so that a tonne of CO_2 means the same in different countries/sectors involved in the carbon market.

For a cap-and-trade system to function correctly, the **registries** must ensure that the emissions and corresponding emission rights (quotas or credits) can be tracked reliably and securely. The registries, which consist of large sets of accounts on the cap-and-trade system, ensure that transfers of ownership of emission rights are recorded. In existing systems, these transfers of ownership take place in real time, meaning that a registry has no *futures* or *forwards* – only spot sales can be recorded there. Registries ensure only that there is an inventory of the quantities traded and they do not contain any information on the allowances price. Their principal function is to ensure that the volume of quotas can be traced, thus guaranteeing the environmental integrity of the system. The quantity indicated in the register provides a reliable record of emission rights at the end of the accounting period or during reconciliation of actual emissions, in addition to emission rights held by participants.

Finally, to guarantee the environmental integrity of the system, that is, to ensure the quantity of emissions is below the cap, the regulator must impose **sanctions** to penalize agents who cannot cover their emissions with an equivalent number of quotas. A system of fines encourages polluters to ensure that their emissions do not exceed the number of quotas in their possession. However, these fines alone are not always enough to guarantee compliance with the environmental target. Suppose that the demand for quotas exceeds supply (in the case of an ambitious environmental policy). The allowance price would rise to reach a balance between supply and demand. If paying a fine removed the obligation to secure allowances covering emissions, the cap might not be respected, because agents would prefer to pay a fine every time that the allowance price was higher than the cost of paying the fine. To avoid this threshold, the legislator could specify that paying a fine did not free agents from their responsibilities: the defaulting entity would also have to buy the missing rights during future compliance periods.

FIGURE 6.1 Constituent elements of an emission-rights market

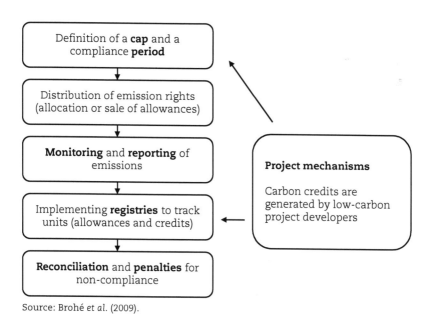

Source: Brohé *et al.* (2009).

Setting up a carbon market

In the early 1990s academic and political networks started to promote the creation of carbon markets (Paterson, 2012). Supporters of the approach generally cite its economic efficiency and flexibility, and the possibility of negotiations or political bargaining between developed and developing countries.

The first article on this subject was published by Michael Grubb in 1989. The article, "The Greenhouse effect: negotiating targets", showed the benefits of this approach's flexibility and political feasibility (Grubb, 1989). In 1992, the first report on this issue by an official UN body discussed the opportunities offered by this type of instrument (UNCTAD, 1992).

Networks of experts played a significant part in enabling these markets to emerge, become accepted and be taken up by political decision-makers, at a national, the European and the UN level. Paterson identifies two groups of technocrats who played a particularly important role in the emergence of carbon markets. Firstly, one can see the influence of British decision-makers, particularly following a change of strategy by the Confederation of British Industry (CBI, the largest employers' union in Britain). In the mid-1990s, British employers adopted a more constructive approach to development of environmental regulations, instead of opposing all regulations head-on or sticking their head in the sand as was previously the case. The British also support the carbon-market system as they want to strengthen the central role of the City of London in capital markets. At the European Commission, various officials have highlighted the theoretical benefits of market mechanisms since the late 1990s, and this body is the second network to play a key role in the creation of carbon markets. Influenced by these technocrats, and also with a view to saving Kyoto, European political decision-makers changed their position radically between 1997 and 2003 (Skjærseth and Wettestad, 2008). One of the most influential European officials has been Jos Delbeke, a Belgian economist who was the director general of the now-defunct DG Climate Action at the European Commission. Works by Jon Birger Skjærseth, Jørgen Wettestad (2008) and Marcel Braun (2009) maintain that DG Environment is acting like a policy entrepreneur at the heart of the political network of the EU. A comparison between the European Commission green paper published in 2000 (EC, 2000), and the directive in 2003 shows that the content and form have remained virtually intact throughout the European decision-making process. The main change, made under pressure from industry, was to move the centre of decision-making for granting quotas from the Commission to the member states (Markussen *et al.*, 2005). European officials could also use the polluter-pays principle to defend the idea of setting up a carbon market. This principle has been one of the key principles of European law since the Single European Act came into force in 1987. Title XIX lays the legal foundation for EU environmental policy: "Community policy shall be based on the precautionary principle and on the principles of preventive action and rectification of damage to the environment at source as a priority" (observing the polluter-pays principle), as per Article 191 of the Treaty on the Functioning of the European Union.

However, the networks of technocrats cannot entirely explain why this instrument was chosen, or why it was accepted by business. Since the Earth Summit (1992), a growing proportion of companies have been convinced that climate change represents a threat (to the insurance sector, for example) or conversely a business opportunity (the renewable energy sector, nuclear energy, the financial and services sector etc.). In 1999, BP created its own internal system for trading emission rights. Shell and Lafarge were also quick to support the principle of emission rights (Braun, 2009).

The first private actors working exclusively with carbon finance and management emerged after Kyoto in 1997. The companies EcoSecurities and Climate Care are two such companies, both based in Oxford in the United Kingdom. EcoSecurities is a pioneer in the development of projects arising from the Kyoto flexibility mechanisms, and Climate Care is particularly active in voluntary offset markets. The International Emissions Trading Association (IETA) is an association founded in 1999 that brings together companies concerned with carbon markets. In 2000, Norwegian entrepreneurs created PointCarbon, now part of Thomson Reuters, which supplies information and advice to actors in the carbon market (companies, traders, brokers, investors etc.). Shortly after 2000, the first big banks set up departments specializing in "carbon assets" (Barclays, Cantor Fitzgerald). Around 2005, other large banks developed their expertise internally (for example, Deutsche Bank, the Société Générale, which notably set up a joint venture with Rhodia under the name Orbeo, as well as ABN Amro and ING in the Netherlands) or by integrating pioneering companies (JP Morgan bought and incorporated Climate Care and EcoSecurities into its company in 2008). In France, the Caisse des Dépôts played an important role through targeted investments and developing technical expertise in the area of registry development and market analysis. Since 2011, many banks have reduced their carbon finance activities as a result of the drop in carbon credit prices. For example, Climate Care became independent after a management buyout. Two events in December 2012 illustrate how seriously the crisis affected the sector: Bluenext, a platform for trading carbon credits and quotas, owned by NYSE Euronext and CDC, closed, and Deutsche Bank ceased its activities in carbon finance.

Kyoto and the carbon market

Cap and trade under Kyoto

The Kyoto Protocol commits the Parties in Annex B, that is, developed countries (members of the OECD or the former Soviet bloc), to a binding objective for reducing or limiting their GHG emissions.

The commitment period was from 2008 to 2012. The objective is measured over several years as there may be significant variations from one year to another, for example due to seasonal or temporary factors.

The Protocol recognizes six gases or families of GHGs: CO_2, CH_4, N_2O, HFCs, PFCs and SF_6. It should be noted that until January 2015, the United Nations still recognized the GWPs provided by the IPCC before the negotiation of Kyoto, that is, values from 1995.

Allocation was made on the basis of historical emissions. This approach favours countries that have historically been more industrialized. Unless recently industrialized countries are authorized to increase their emissions significantly by comparison to historical baselines, this method of allocation would be very unfavourable to developing countries. The quotas allocated are known as **Assigned Amount Units**. These **AAUs** represent a metric tonne of CO_2 equivalent, and they are tradable by virtue of Article 17 of the Kyoto Protocol. The trading registry itself is called the ITL (International Transaction Log), and it comprises an accounting tool that enables the observance of the rules established by the Kyoto Protocol to be verified.

Climate targets vary significantly depending on the country: Iceland, for example, was authorized to increase its emissions by 10% between 1990 and the 2008/2012 average; the United States was committed to a 7% reduction, the EU to 8% and so forth. Intuitively, one might think that countries committed to a greater reduction would have more difficulty reaching their objectives than those countries authorized to make a smaller reduction or *a fortiori* an increase. However, this reasoning is erroneous, as the actual context determines how hard it is to reach targets. For example, a country with a lot of ageing coal-fired power stations in 1990 (such as the United Kingdom or Germany) could go a long way towards meeting its reduction targets by setting up more efficient power stations. Similarly, a country where the service sector has expanded since 1990 would find it easier than a country where industrialization has increased since 1990.

There are two exceptions where the reference year chosen may not be 1990. Countries in transition to a market economy, which saw their emissions drop spectacularly in 1990, may use a year or historical reference period prior to 1990 in order to meet their commitments. Secondly, it is possible to choose 1995 as a reference year for calculating emissions of HFCs, PFCs and SF_6. These points of flexibility, which are the fruit of political negotiations, increase the reference level and reduce the environmental target. Indeed, following the Montreal Protocol (1987), a range of substances that damage the ozone layer (including CFCs) were progressively banned or replaced with substances that are not damaging to the ozone layer but have high global–warming potential, particularly HFCs. This explains the significant increase in these emissions between 1990 and 1995.

GHG emissions from motor fuels used for international air and maritime transport are not taken into account in the framework of the Kyoto Protocol due to compatibility difficulties and political deadlock at specialized UN institutions (the International Civil Aviation Organization and the International Maritime Organization). Fifteen years later little progress has been made, and the EU has decided to include the aviation sector in the European carbon market, angering the United States, China and India. It is worth noting that in the context of road transport, the country responsible for emissions is the country that sold the fuel, which significantly inflates emissions for a country with lower levels of fuel taxation like Luxembourg, due to "petrol pump tourism".

In practice, demand for allowances was determined by the actual emissions of Annex B Parties during the period 2008 – 2012. Supply is determined by the cap, but it is increased by the volume of credits generated through the Clean Development Mechanism (see following section).

Russia and Ukraine had a net surplus (5.8 billion unused allowances for Russia and 2.6 billion for Ukraine) during the five years of the compliance period. This over-allocation to countries in transition, sometimes termed "hot air", has often been decried by critics of the international carbon market. In 1997, when these countries' targets were defined, their emissions were already well below the levels outlined in their commitments, which meant that they received millions of excess AAUs.

Meanwhile, many western European countries have seen their emissions fall dramatically since 1990, particularly in the last few years as a result the economic crisis. Consequently, the EU had an overall surplus, but there were significant variations between member states. Among the countries that have ratified the Protocol, Canada and Japan are the countries with the highest allowances deficits. Canada's deficit is by far the highest. Between 1990 and 2009, their emissions increased by 17% (despite a marked slowdown in the economy from the end of 2008). The main reason for this increase is the use of tar sands by the oil industry in Alberta.

To meet its commitments, Canada would have had to reduce its emissions by 6% over the same period. As a result, its deficit over the compliance period is around 700 million quotas. This failure to reach the target and the lack of political will to invest in purchasing quotas or credits abroad pushed Canada to withdraw from the Kyoto Protocol in December 2011. In doing so, Canada risked being ostracized by the international community. Although there are no financial sanctions for actors who do not meet their commitments (unlike the European carbon market), Canada risks having to meet more severe targets if it wants to join a future UNFCCC protocol. The primary sanction included in the Kyoto Protocol's framework (Marrakesh Accords) is the reduction of allocation for a second period of commitment by 1.3 times the excess emission not covered in the first period.

In addition to surplus allowances, there are credits generated by project mechanisms. The Clean Development Mechanism has released 1.2 billion credits, and joint implementation has released more than 350 million credits during the first compliance period (Shishlov *et al.*, 2012).

This huge surplus of allowances remains a source of concern for countries participating in the second commitment period. In 2015, the Kyoto Protocol suffered from a surplus of 11 gigatonnes inherited from the first commitment period. This surplus seriously undermines its environmental effectiveness. During the Paris conference in December 2015, five countries (Denmark, Germany, the Netherlands, Sweden and the United Kingdom) announced that they would cancel a portion of their pre-2020 surplus units (AAUs) under the Kyoto Protocol. With the goal of supporting an ambitious Paris climate agreement, they cancelled 634.9 million surplus units from the Kyoto Protocol's first commitment period.

Note that under the current rules, the surplus AAUs under the Kyoto Protocol cannot be used after 2020. However, other Kyoto Protocol units, such as pre-2020 carbon offsets, could still severely undermine the environmental integrity of the Paris climate agreement if Parties allow them to be carried over.

The Clean Development Mechanism

Sometimes dubbed the "Kyoto surprise", as it was added right at the end of negotiations, the Clean Development Mechanism (CDM) is one of three flexibility mechanisms established by the Kyoto Protocol, together with (international) emission trading and joint implementation. It consists of investment by an Annex I country in a non-Annex I country to reduce GHG emissions and work towards sustainable development. For each tonne of CO_2 reduced or absorbed through the project, the investor receives a **Certified Emission Reduction (CER)**. This mechanism to generate credits can be used to cover emissions by Annex I Parties, and it is defined as an **offset mechanism**. Although the objective of contributing to sustainable development is explicitly mentioned, it is not clearly defined.

In theory, this mechanism should favour transfer of technologies and capital within a sustainable development perspective. We can see that in practice, capital is concentrated in certain technologies and certain countries, and little capital is mobilized for the least advanced countries and the most complex technologies.

It was the **Marrakesh Accords** (COP 7 in 2001) that defined the key practical details of the CDM. These accords make prevision for establishing the **CDM Executive Board** and specify the different stages leading up to release of carbon credits.

The Executive Board is in charge of implementing the practical details and procedures of the CDM. The Board is made up of ten members, and it is responsible to the COP for implementation of the CDM. Specifically, the Executive Board is responsible for approving methodologies, monitoring plans and accrediting operational entities (auditors). It is also responsible for developing and maintaining a registry of projects. It is assisted in its work by designated operational entities (DOEs). These entities are responsible for validation, verification and certification of CDM projects. The Board plays a fundamental role and it has been the subject of much criticism in evaluations to improve the functioning of the mechanism. A key criticism has been the lack of transparency in decision-making on project registration and political games among members to gain influence.

Box 6.1   Additionality of offset projects

Additionality is the basic criterion for an offset project to be recognized. To work this criterion, project developers need to start with a baseline scenario, to show that their project causes a reduction in GHG emissions that would not happen without it. It is the reduction in GHGs between the baseline scenario and the scenario with the project that means the developer is entitled to credits. To check the environmental additionality, it is necessary to verify the additionality of investment and technology. Additionality of investment means that a CDM project should bring in supplementary investment compared to the baseline scenario. This supplementary investment is intended to finance emission reductions. Technological additionality is the principle that an offset project should involve North-South technology transfer. This means that the technology used to reduce GHG emissions would not be present in the project areas due to the cost or complexity of importing it unless there was additional support. Finally, economic additionality implies that the capital contributed by developed countries could not be replaced by official development aid (ODA).

One weak point of the system that it uses the concept of additionality to determine the volume of emissions avoided and the number of credits to which a project is entitled. As an offset mechanism, the CDM does not increase the volume of emission reductions. The system is only environmentally neutral only if the emissions are really additional compared to the level of emissions that would take place without the political intervention. Gillenwater (2012) criticizes the usual imprecise and circular definition of additionality which states that "a project is additional if it creates emission reductions compared to a scenario without the project". Gillenwater believes that this lack of precision leaves evaluation of additionality vulnerable to political decisions and *ad hoc* justifications. The tools used by the United Nations are not currently capable of distinguishing additional projects from non-additional projects. For example, Haya (2009) showed that most hydro-electric power stations in China receive carbon credits, although similar projects have previously been constructed without carbon financing and the decision to construct new projects had often been made before creation of the CDM.

The CDM has been the subject of much criticism. This is essentially because it is the principal existing mechanism for fighting GHG emissions in developing countries. The aim of the mechanism is to help developing countries through technological and financial transfers for low-carbon development. Many doubt the mechanism's capacity to achieve this ambitious goal. The following summary presents some of the most common criticisms made when evaluating the success of CDM projects.

Procedures for project validation, verification and certification entail transaction costs that are often insurmountable barriers for project developers. This criticism is not new, and the Marrakesh Accords included a simplified procedure for small-scale projects to address the issue. Despite this measure, transaction costs remain high – varying between estimated figures of 50 EUR and 100,000 EUR, and also affecting smaller projects.

Another common criticism concerns the geographical origin of credits (and therefore the transfer of capital). More than 90% of credits supplied come from four countries (China, India, South Korea and Brazil). Just 1% of credits supplied in early 2012 came from Africa, of which 95% came from Egypt or South Africa. In practice, this has seriously eroded the image of a mechanism that would contribute to development in Africa and less advanced countries.

The CDM Executive Board has worked to improve the CDM, in response to demands from Parties and reflecting stakeholders' contributions, reducing complexity and transaction costs, introducing new and innovative approaches for quantification of mitigation outcomes, and broadening the regional distribution of projects.

The programme of activities (PoA) approach introduced in 2007 is an example of improvement. These programmes involve including small emission-reduction projects by actor, by sector and by region, which were not previously covered by the traditional CDM approach. This can encourage investments such as introducing new, more energy-efficient equipment or small-scale measures in the area of renewable energies, such as solar water heaters or domestic biogas. This programme was successful in encouraging smaller projects, especially in underrepresented countries, by allowing an unlimited number of activities to be administered under a single programme. The approach was well received by project developers and stakeholders, with more than 265 PoAs registered in 75 countries.

The technologies that generate credits have also been criticized (Olsen, 2007; Sutter and Parreño, 2007). Specifically, the projects that provide most credits involve the most profitable activities. This is desirable in economic and environmental terms, and it is the foundation of implementing an emission-rights market.

However, the CDM was designed to embody the principles of sustainable development by adding a social dimension to these two objectives. Although most projects implemented have involved energy (small hydro-electric power stations, wind farms, energy efficiency etc.) or waste (waste recycling, methane recovery etc.), the picture is very different with regard to the number of credits generated. Indeed, industrial projects for destruction of HFCs or N_2O and large dams have taken the lion's share (75% of credits supplied at the start of 2012 came from one of these three categories). Just 6.8% of credits came from wind farms and 0.2% from solar energy projects.

For example, the 22 HFC destruction projects (0.6% of the total number of projects recorded) generated more than 45% of the credits issued before 2012. This type of project has been so successful because of the exceptional return on investment (compatibility of CO_2-equivalents means that destroying 1 tonne of HFC-23 equates to the same as preventing the emission of 11,700 tonnes of CO_2) and because it is easy to prove additionality. Indeed, without credits there is no incentive to destroy HFCs (as the process neither saves energy nor produces clean energy or biofuels) so the activity must be additional, as credits are the only source of revenue for the project. The CDM has also been accused of encouraging developing countries not to adopt regulations to eliminate HFCs so as not to damage this capital transfer. Indeed, if China had legislated to make HFC destruction obligatory these destruction projects could no longer be considered additional, so they could no longer generate credits. More worryingly, it has been shown that companies could gain more revenues from avoiding HFC-23 emissions (a co-product from the production of commercial coolant HFC-22) than from their core business (Wara, 2007). As a result, many sites have been accused of artificially inflating their coolant production with the sole aim of generating more HFC-23 and therefore receiving more credits for destroying this gas. For this reason, the Chinese government taxes these projects heavily. Furthermore this type of credit is no longer recognized by the EU ETS, and the methodology has been reviewed by the Executive Board of the CDM.

In only 12 years of functioning, the CDM has registered almost 7,300 projects in 89 countries, and it has delivered more than 1.38 billion tonnes of emission reductions. Despite these impressive outputs, the CDM is facing the most serious crisis of its brief existence (Brohé, 2014). Demand from traditional markets (especially the EU ETS) has contracted severely, with the spot price of a secondary CER plummeting from more than 20 EUR in 2008 to around 0.40 EUR in 2016. As a consequence, investment in new CDM projects has been almost non-existent since 2013, and significant haemorrhaging of private sector capacity is occurring.

In 2016, one of the CDM's remaining unique strengths is its utility for monitoring, reporting and verifying action to reduce GHG emissions. As such it can represent an opportunity for funding agencies – within and outside the UNFCCC – that wish to finance mitigation projects on the basis of performance, or results-based financing.

Joint implementation

The rules and practical details of joint implementation, the flexibility mechanism introduced in the Kyoto Protocol (Article 6), were also specified in the Marrakesh Accords in November 2001. Projects undertaken in the framework of joint implementation should be carried out in an Annex I country. As these countries have binding emission targets, the emission reductions achieved as part of a joint implementation project transform the assigned amounts (UUA) in the host country to an equivalent number of **Emission Reduction Units (ERUs)**. These **ERUs** are then transferred to the investor country. This is a fundamental difference with the CDM: in this case no additional emission rights are created. The AAUs must be cancelled to avoid double counting. Indeed, if Germany (or a German company) finances a project for 80,000 tonnes of CO_2 emission reductions in Russia, it is entitled to 80,000 ERUs. Russia requires fewer AAUs to meet its target, thanks to the emission reduction financed by Germany. The possible surplus could then be sold again through AAU trading, which would then cause double counting of the avoided emissions.

As with the CDM, one of the basic criteria to be respected is additionality. However, the risk of the registration of non additional projects is lower as the credits that are created automatically cancel the AAUs. This means it is bad for the host country to approve a non-additional project, because it loses AAUs. Environmental integrity is therefore guaranteed as the credits generated are not added to existing emission rights, but rather they replace AAUs. With joint implementation, it is in the host country's interests to ensure that projects generate effective emission reductions, which are measurable in the long term. If this is not the case, the host country risks transferring more ERUs than the volume of emissions actually generated, weakening its own capacity to meet its reduction target. To put it another way, the host country has no *a priori* interest in the benchmark threshold being too high. This is not the case with the CDM, in which Parties have a shared interest in figures turning out a certain way. For this reason it is vital that impartial checks are carried out on the operational entities and the Executive Board to guarantee the credibility of the system. However, in practice, as AAUs are less fluid and thus less onerous than ERUs, there is a temptation for countries with a surplus of AAUs to grant environmental projects an inflated number of ERUs.

The procedure for developing a project in the framework of joint implementation was designed to be quick and simple, and the determining factor is agreement between the parties involved. The host country is then free to apply the provisions it wishes in order to approve the project and transfer the ERUs. However, this procedure can only be used if the host country meets all the eligibility criteria and has indeed adopted the guidelines for project recognition.

If the host country does not meet all the criteria, and would not be eligible, the Marrakesh Accords offer a second, different procedure, known as "Track 2". This procedure copies the life-cycle of a CDM project. Projects developed within the framework of joint implementation Track 2 are supervised by the JI Supervisory Committee (JISC), an international body comparable to the Executive Board of the CDM. Note that even if the host country meets all the eligibility criteria, a joint implementation project may still be developed through Track 2 on a voluntary basis.

For JI track 2, an Accredited Independent Entity (IEA) plays the role of the operational entity. Before 2008, only Track 2 projects were underway. NGOs and independent carbon-market observers feared that Track 1 could not "green" the AAU surplus of countries in transition, like Russia or Ukraine, because of the significant role played by the host country in evaluation of additionality and determining emissions avoided.

Russia and Ukraine together generate 80% of ERUs. Just less than 20% is generated by EU countries that have used this mechanism – mainly countries in Eastern Europe, as well as France and Germany. New Zealand also has some joint implementation projects (Shishlov et al., 2012).

In total, 548 Track 1 and 52 Track 2 projects were approved and over 871 million emission reduction units issued in the period 2006 – 2015.

The European carbon market

Learning by doing

The introduction of a European carbon tax failed in the 1990s, while national carbon markets continued to develop in the United Kingdom and Denmark. In this context the European Commission started to investigate the possibility of setting up a European carbon market covering industrial emissions in the late 1990s (Brohé, 2008).

Directive 2003/87/CE (EU, 2003b) set up an **EU Emissions Trading System (EU ETS)** in January 2005. It is the most ambitious instrument in European climate policy. Drafted before Kyoto came into force, the directive does not make the functioning of the market dependent on the future of the Protocol. At times when the future of Kyoto is uncertain, the absence of a direct link enables the system to continue.

The EU ETS was the first functioning international trading system for CO_2-emission rights. The system covers more than 10,000 facilities in the energy and industrial sectors, which are collectively responsible for nearly half of CO_2 emissions in the EU and 40% of all GHG emissions. Airlines have been included in this system since 1 January 2012. Although all flights with an arrival or departure point in the EU were originally to be covered, in November 2012 the Commission proposed to exempt flights going outside European territory following pressure from important parties, particularly the United States and China.

The EU ETS is a cap-and-trade system, that is, the overall permissible level of emissions is capped, but participants are authorized to buy and sell emission allowances to meet their needs up to this limit. The rights allocated within the EU ETS are EU allowances (**EUA**). All transactions can be traced in a central registry, known as the

European Union Transaction Log (EUTL). This registry traces allocations, transactions and withdrawal of allowances in the EU ETS.

Before the start of the two first trading periods (2005 – 2007 and 2008 – 2012), member states established **National Allocation Plans** (NAPs), which were then validated by the Commission. These plans set the number of emission allowances attributed to each facility in a member state's territory, also making provision for a reserve for new entrants (companies created after implementation of the system) and (national) limits on import of CER/ERU credits.

The emission targets (the cap) for each participant is therefore decided at a national level on a site-by-site basis. This approach can lead to protectionism and distort competition. For example, three electric power stations using the same technology receive a different number of allowances depending on whether they are located in France, Germany or Poland. With this approach, the number of allowances offered by member states becomes a decisive criterion when choosing where to set up a facility with high emissions. To deal with this alleged distortion of competition by the 27 national allocation plans, the system was replaced by a single cap for the entire EU for the third trading period (2013 – 2020). The cap is scaled down following a linear trend, with an annual reduction of 1.74%, continuing beyond the term of the third trading period (2013 – 2020). Another major change since the start of the third period is that quotas are allocated on the basis of harmonized rules and a significant proportion of these quotas are subject to bidding. Some of the revenue from bidding is redistributed from the member states with a high per capita income to the poorest member states.

At the beginning of the third trading period (2013 – 2020), manufacturing industry received 80% of its allowances for free. This proportion decreases gradually each year to 30% in 2020. Power generators since 2013 in principle do not receive any free allowances, but have to buy them. However, some free allowances are available to modernize the power sector in some member states. Airlines continue to receive the large majority of their allowances for free in the period 2013 – 2020. Benchmarks were set up for free allocations. These benchmarks, based on historical data, favour managers who have already taken measures to reduce GHG emissions. Note that although free allocations are made on the basis of benchmarks, the volume of allowances granted is still linked to historical production volumes, which in turn will benefit actors who have significantly reduced their production volumes since the economic crisis.

In theory, free allocation was to disappear after 2020. However, an exception has been made for facilities in sectors with a risk of carbon leakage, around 50 sectors. Free allowances will still be attributed but in accordance with common rules, so that all companies in the EU with identical or similar activities are subject to the same rules. More flexible rules to better align the amount of free allowances with production figures have been introduced for the fourth phase (2021 – 2030). In total, it is expected that around 6.3 billion allowances will be allocated for free to companies over the period 2021 – 2030.

Companies participating must submit allowances covering their emissions by 30 April of the following year. However, if a company does not supply enough emission rights to cover its actual emissions during the previous year, it must pay a fine of 100 EUR per tonne of CO_2-equivalent emitted that is not covered by emission rights (allowances or credits). This fine does not release the company from the responsibility to acquire the necessary emission rights on the market, and it does not represent a cap price. In addition to paying the fine, a company in default must buy the missing allowances during the following year.

Fraud and regulation of the European carbon market

In 2009 the authorities were alerted to the fact that the ETS was the target of VAT fraud. In this case, the fraudsters set up an account in one country, then bought allowances from a vendor in another country. They did not pay the VAT as the community rules exempt cross-border allowances sales from VAT. The fraudsters then sold the allowances on in national transactions and added VAT. However, instead of submitting the VAT collected to the state, they pocketed the difference and disappeared. Also, if the final buyer was a company, it requested a refund from the state for the VAT it had paid. The state then reimbursed taxes that it had never received. This type of VAT fraud could theoretically take place in any market that, like the carbon market, allows quick sale of high-value merchandise, with the VAT to be declared by the vendor. However, the abstract nature of the merchandise and the large volumes traded in allowances transactions enabled fraud on a huge scale, probably costing 1.6 billion EUR in France and around 5 billion EUR across the EU. Immediately after this was discovered, the authorities suspended VAT on carbon-market transactions and some arrests were made. Cyber criminals have also managed to steal some allowances by attacking several national emission registries (phishing).

These cases of fraud raise the important question of how to regulate this financial market. The Prada Commission (Commission Prada 2010) recommended consolidation of the current structure of organized markets (platforms) and mutual agreement on better market supervision. This supervision is particularly important as the CO_2 market is principally a market for derivatives. A framework to prevent and punish market abuses, geared to the specific characteristics of the CO_2 market, should be set up to limit the risk of rate manipulation and insider dealing (Prada Commission, 2010).

Recognition of the Kyoto project mechanisms

The directive 2004/101/EC (EU, 2004b), widely known as the linking directive, links the allowances exchanged under the EU ETS and the credits obtained via the Kyoto flexibility mechanisms. This directive makes provision for equality between three types of emission right: an emission right obtained by carrying out a project using joint implementation, an emission credit obtained by carrying out a project using the Clean Development Mechanism, and an emission right obtained in the framework of the EU ETS. European law specifies certain additional conditions for use of CERs and ERUs in the EU ETS. For example, credits generated by projects linked to land use, land-use change and forestry are not authorized in the EU ETS. Credits generated by hydroelectric generation projects with a capacity of more than 20 MW must respect the relevant international guidelines and criteria, and those contained in the final report of the World Commission on Dams published in 2000. CER credits from projects for destruction of HFC-23 and N_2O have been banned since 2013.

With 80% of the global volume in 2015, the European carbon market is by far the largest carbon market. Different quality criteria added by the EU therefore have an impact on the fungibility of allowances. Indeed, a credit from a large hydroelectric project that does not respect the additional criteria of the World Commission on Dams or an HFC-23 project loses value compared to others due to the weaker demand for their credits.

There has been a price difference between CERs (see Figure 10) and EUAs since the start of 2012, as the quantity of carbon credits that can be imported into the system is limited for each company. Most large companies have used their full entitlement, thereby drastically reducing the demand and therefore the prices for Kyoto credits.

The linking directive has had a significant impact on the EU ETS. The projected surplus in the EU ETS (about 900 million allowances) is in fact very close to the additional inflow of emission reduction credits from CDM and JI projects. In other words, without the linking directive, the imbalance between demand and supply would be much smaller and prices would certainly be much higher. Not only was this inflow detrimental to the economics of the market, it has also raised serious environmental concerns as the additionality of many of these credits has been questioned. The vast majority of CDM credits surrendered under the EU ETS since 2008 were created by projects that are now excluded from use under the EU ETS because of integrity concerns (Brohé *et al.*, 2016).

Falling prices and calls for intervention

The start of the third period (2013 – 2020) is a time of transition, when the system must still prove that it can weather the economic crisis that has battered Europe since late 2008. Indeed, the unfavourable economic situation, and to a lesser extent interaction with other climate and energy policies led to much lower emission levels than predicted when the NAPs were set up (before 2013) and the linear factor for annual reduction was established (after 2013). As a result, demand for allowances has generally been below supply, causing prices to drop from 29 EUR in June 2008 to just 2 EUR in April 2013. In August 2016 prices were slightly below 5 EUR. This may be a logical reaction by the market, but it has still led to calls for intervention to support prices (Sartor, 2012). The European Parliament Environment Committee proposed withdrawing up to 1.4 billion allowances from the market in the course of phase 3. The reason given was "to compensate for the implementation of the directive on energy efficiency" and to "re-establish the price mechanism at the levels envisaged by the impact analysis". There was also a proposal to increase the annual reduction factor for allowances supply from 1.74% to 2.25% from 2014, to be consistent with the target of 80% emission reductions in the EU by 2050. The market reacted with a strong initial rebound, and the price of EUAs climbed by 30% on the day of the announcement, only to fall back when it became clear that there were political difficulties with adoption of these measures.

It is important for market actors that any EU intervention in the market is for a limited period and justified, so it does not create too much uncertainty. If the reasons for an intervention are widely understood and accepted, and the number of allowances removed is consistent with this reasoning, an intervention could increase confidence in the authorities' capacity to manage unexpected situations and protect the interests of investors in low-carbon technologies and procedures (Sartor, 2012).

FIGURE 6.2   Changes in prices for European Union Allowances (EUAs) and CERs between 2008 and 2016

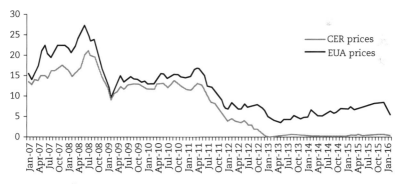

Source: CDC Climat.

Other carbon markets

Markets operating in Australia and New Zealand

The State of New South Wales in Australia implemented the first obligatory regional cap-and-trade system for emission allowances in 2003 (Delbosc and Goubet, 2011). Since July 2012, a federal system has established a fixed carbon price for a period of three years. From 2015, the mechanism will change to a cap-and-trade system linked to the EU ETS. Australia's neighbour, New Zealand, launched its own system in 2008. This country has an emission profile very similar to that of a developing country, as half of emissions come from the agricultural sector and 25% are offset by forestry activities. A unique feature of the New Zealand system among carbon markets is that it has covered emissions from the forestry sector since its creation and as from 2015 it includes agriculture, which currently has reporting but not surrender obligations (Delbosc and Goubet, 2011). The other systems generally considered that there were insufficient guarantees for the permanence of forestry activities for carbon credits to be granted. Originally, the New Zealand ETS was designed to provide unlimited access to international Kyoto units. However as of 1 June 2015, the government effectively restricted it to a domestic-only system. This may again change in the future as the government has indicated access to international markets as one of its priorities.

Voluntary system in Japan, pilots in China and domestic offsetting in Taiwan

Japan has implemented two voluntary cap systems. The first, the *Keidanren Voluntary Action Plan* (K-VAP) was launched by the main Japanese employers' association, and the second, the *Japanese Voluntary Emissions Trading Scheme* (J-VETS), was developed by the Environment Ministry. The *Voluntary Carbon Credit Trading Scheme* is a voluntary initiative that attempts to bring together and extend the area covered by the two systems mentioned above (Delbosc and Goubet, 2011). Some Japanese municipalities, particularly Tokyo, have also launched local systems. The various Japanese systems authorize unlimited use of credits from the Kyoto Protocol. This means that Japan is one of the main buyers of international credits, particularly Chinese CERs. The Japanese authorities regularly criticize these transfers of funds to China, partly explaining Japan's opposition to implementation of a successor to the Kyoto Protocol. Due to opposition from industry and Japan's difficulties with its Kyoto target, it is unlikely that an obligatory system will be implemented in the short or medium term.

In China, the 12th Five-Year Plan (2011–2015) provided for six new pilot projects in different provinces or municipalities (Chiquet, 2015). This pilot phase has been running since 2013 in Beijing, Shanghai, Tianjin, Chongqing and Shenzhen, as well as in Guangdong and Hubei Provinces. China's proposed ETS promises to expand global emission coverage from 9% to 16%. The program is intended to regulate a variety of sectors, including power generation, petroleum refining, chemical and material production, metal manufacturing and aviation. There are challenges that implementing this ETS may face, however. It may prove difficult to design a uniform set of rules and regulations for these sectors. Also, it may be challenging to ensure and enforce compliance of these rules and regulations, especially when considering how rules for monitoring, reporting and verifying (MRV) will be implemented. In addition, special care will need to be taken to ensure that over-allocation is avoided. Since China is the world's largest emitter, the Chinese ETS will have major impacts on both global climate policy and global carbon markets. It is estimated that the cap will lead to the allocation of 4 billion emission rights (tonnes). In comparison, the EU ETS allocates about 2 billion emission rights.

Local initiatives in the United States and Canada

In the United States, Congress is opposed to all forms of GHG emission regulations. However, this lack of action at federal level should not detract from the numerous initiatives launched by states or groups of states (Brohé *et al.*, 2009). The situation changed largely after 3 August 2015 when President Obama unveiled the final version of the Clean Power Plan (CPP). The main goal of the plan is to reduce emissions by electrical-power generation by 32% within 25 years relative to 2005 levels. Other than that the plan largely promotes the use of renewable energy and energy conservation. Individual states will need to meet the proposed standards in respect to the carbon emissions and submit their own reduction plans by September 2016. If a plan is not submitted, ETA will impose its own plan on the state.

For many of the states regulators, carbon trading is indeed a very appealing way to comply with the CPP regulations. While the CPP itself does not specifically reference the market oversight, it provides for transparency, accuracy and consistency to help and ensure a fair and functional market.

There are currently two existing carbon-trading programs in the United States – the Regional Greenhouse Gas Initiative (RGGI) and California's Cap-and-Trade Program. Both distribute allowances via auctions, and the overseeing agencies are responsible for determining the number of allowances traded. RGGI, which was the first mandatory market-based program in the United States, at one time had 10 US states participating, including Connecticut, Delaware, Maine, Maryland, Massachusetts, New Hampshire, New Jersey, New York, Rhode Island and Vermont. RGGI reduces CO_2

emissions by capping the emissions of large electricity generators and holding quarterly auctions for allowances.

California's Cap-and-Trade Program, which was a product of the greatly diminished Western Climate Initiative (WCI), is the only other active US emissions market. It blends a market-based approach with some unique features, including the permission of offsets and international trading with Quebec. Allowances are acquired by regulated entities by both free allocation and quarterly auctions. Regulated emissions follow the six regulated gases included in the Kyoto Protocol, but they also include a few other fluoridated gases. In 2015, the program was expanded to include the regulation of fuel distributors. This cap-and-trade market is an integral part of the California Global Warming Solutions Act of 2006 (AB 32), which aims to reduce GHG emissions in the state by 25% by the year 2020.

Canadian provinces have proven to be important partners of US states in the effort to develop regional cap-and-trade markets. The WCI counts three other Canadian provinces as future participants of the international trading program to which Quebec and California already participate: Ontario, Manitoba and British Columbia. Building upon these efforts to diffuse carbon markets, President Obama and Prime Minister Trudeau met in March 2016 to discuss climate change. At the occasion, Prime Minister Trudeau proposed a pan-Canadian plan to price carbon, which would make Canada the first North American country to institute an economy-wide price for carbon.

A shift from a top-down policy architecture leads to more fragmentation

Overall, despite the surplus and consequent price crash observed on the two main carbon markets (EU ETS and CDM), emissions trading is gaining traction as the preferred type of carbon-pricing policy instrument for many governments around the world. The number of emissions trading schemes (ETSs) has more than tripled since 2012, going from five to 17. But the carbon markets of 2016 look very different from those of a decade ago. The world has shifted from the top-down policy architecture, initiated by the Kyoto Protocol, into a bottom-up architecture under the newly adopted Paris Agreement where governments set targets at a national level, so-called "Nationally Determined Contributions" (NDCs), and adopt various policy approaches, not all of which are market-based.

The current fragmentation into national and regional carbon markets may pose a challenge for creating a more globally connected carbon market in the future.

Voluntary offsetting for GHG emissions

Simple principle, risk of greenwashing

Voluntary offsetting for GHG emissions is based on a fairly simple approach. Firstly, it is clear that an entity can offset emissions only if it is aware of them. Enterprises that want to offset some of their GHG emissions must first carry out a calculation. When this approach relates to a product or service, the company follows specification PAS 2050, ISO/TS 14067 or the GHG Protocol for products. Then, to avoid accusations of greenwashing, the rules of best practice state that any offset approach should be linked to efforts to reduce emissions internally. The final stage, involving compensation, involves selecting and financially supporting projects that reduce GHG emissions. In practice, these projects are developed in the framework of the CDM, or developed with other standards inspired by the UN framework. The three main voluntary standards are the Verified Carbon Standard, set up by the IETA and the World Economic Forum, the Gold Standard, started by various environmental organizations such as the WWF, and the Climate Action Reserve (CAR), which was built upon the California Registry's knowledge and expertise in GHG accounting.

More than 25 years of voluntary experience

The first company to have used offsetting was the US power company AES Corporation (Bellassen and Leguet, 2007). In 1989 this company decided to finance an agro-forestry project in Guatemala by investing $2 million in the planting of 50 million trees. The goal was to offset emissions resulting from a new power plant built by the group in Connecticut. Since 1991, a German charity called Primaklima has offered companies an offsetting service. In 1997, two British companies (The Carbon Neutral Company and Climate Care) and two associations from the United States and Australia (the National Carbon Offset Coalition and Greenfleet), also entered this market. These five offsetting operators were then joined by three or four new entrants per year. Then in 2006 large banking groups, such as JP Morgan, entered the voluntary market (Kebe *et al.*, 2011) but they left in 2009 because of the financial crisis and plunge in the price of carbon credits.

Transactions are carried out directly between a project developer and the end client only in the case of very large volumes (generally several tens of thousands of tonnes per year) following a call for tender with support provided by consultants and brokers. Specialist suppliers sell retail quantities of voluntary carbon offsetting through transactions ranging from several thousand tonnes for large companies to a few tonnes bought by a private individual in order to offset a flight. Specialist retailers with accounts in the carbon credit registries are often responsible for cancelling credits for the end clients, to prove purchase and guarantee that they will not be sold on.

Growing volumes, declining prices

While the number of credits sold on voluntary offsetting markets quadrupled between 2006 and 2010 (Peters-Stanley et al., 2011), the growth has been far less sustained between 2011 and 2013. From 2014 to 2016, the number of credits sold increased again as more and more companies have been calculating their carbon footprint, for instance to report to the CDP, and using carbon pricing and off-setting as part of their climate strategy. In 2015, the volume of voluntary offset transactions increased by 10% as buyers contracted 84.1 $MtCO_2e$ (Hamrick et al., 2016). However, total market value fell 7% to $278 million, a result of the global average price dropping 14%.

Meanwhile, retirements reached a record 39.5 $MtCO_2e$ in 2015, a 23% increase over 2014. Retirements are a better indication of the final-end user demand for voluntary offsets, since offsets cannot be resold after they are retired on a registry. About two-thirds of the tonnes retired in 2015 were developed under VCS. This was followed by the Gold Standard (20% of 2015 retirements) and the CAR (11%). Prices remain highly variable, with the lowest recorded transaction at $0.1/tonne and the highest reported transaction at $44.8/tonne.

Buyer preferences for particular project types, standards, offset ages (also called vintages) and locations continue to remain influential determinants for price. Offsets from wind beat out Reducing Emissions from Deforestation and forest Degradation (REDD+) as the most sought-after project type in 2015, transacting 12.7 MtCO2. The most supply and demand of any country originated from the United States (15.4 $MtCO_2e$). Buyers also demanded significant volumes of emissions reductions from India (6.6 MtCO2), Indonesia (4.6 MtCO$_2$e), Turkey (3.1 MtCO$_2$e), Kenya (3.1 MtCO$_2$e) and Brazil (3.1 MtCO$_2$e).

Issues and standardization for consumers

Some critics believe that offset mechanisms divert attention by holding out the prospect of an easy climate solution without structural or behavioural change. These detractors think that companies or individuals will continue to act as before, while simply paying a few Euros more. Indeed, they claim that this is a new form of indulgence. In contrast, supporters of the offsetting idea maintain it plays a triple role, and should go hand-in-hand with emission reductions.

Firstly, offsetting is a tool for awareness-raising and education concerning everybody's climate impact. Thanks to this mechanism, individuals and companies are currently aware of their impact, and they are now prepared to act. Secondly, by putting a price on pollution, offsetting acts as a tool for internalization. This argument is often not borne out by the facts for voluntary approaches. However, if an individual takes into account the price of offsetting a short-haul flight, it might become more financially attractive to choose an alternative that causes less pollution (a train). Thirdly, the money invested supports projects that reduce emissions, generally in developing countries. From this perspective, offsetting can be viewed as a lever for climate action.

In practice, offsetting may fit with each of these visions, depending on how it is implemented. Some operators spend a lot of time on auditing and recommendations for reductions, whereas others ignore this section of the process entirely. Some choose their projects carefully to avoid perverse effects and ensure additionality, whereas others gain credits from profitable activities that were already being carried out.

One of the issues regarding voluntary offsetting is that there are no official standards for calculating emissions. While there are guidelines to allow countries that have ratified Kyoto or companies involved in the EU ETS to estimate their emissions accurately, no such rules are defined in voluntary offsetting. If some offsets providers are more scientifically up to date or more accountable and use the latest available IPCC emissions factors, others use inaccurate figures, sometimes in good faith, due to a lack of knowledge.

For example, to estimate emissions from electricity companies it is necessary to know the name of the power supplier and its carbon intensity (usually expressed in gCO_2/kWh delivered to the network). Often offsetting providers use an average rate that is not always representative of the actual context.

These inaccuracies can be even greater when a company calculates its global carbon emissions. A company carbon footprint varies according to the chosen scope. Assumptions also have to be made about the greenhouse gases included (only CO_2 or the six GHGs recognized by Kyoto?), the geographical area (e.g., should travels abroad be taken into account?), the company boundaries (inclusion or exclusion of subsidiaries). These problems explain why CO_2 emissions vary from one carbon audit to another and why independent consultants are often contracted.

Companies in the carbon offsetting sector are faced with the same dilemma as suppliers of organic or fair-trade products. On the one hand, they want to supply "ethical" products, with strong sustainable development guarantees. On the other hand, their prices and costs (including audit costs) should remain reasonable. Finally, a single word or expression – be it "organic", "fair trade" or "carbon neutral" – covers a range of efforts, which may be misleading to the consumer. In the case of organic food, much is at stake financially, so EU regulations were produced on the use of the term in 1991, while fair-trade certification is still organized by the charity sector.

Clearly, when implementing effective measures to mitigate climate change, the ability to differentiate between real and false claims of carbon neutrality is critical. Today, PAS 2060 is the only specification drafted by a standardization body that addresses this issue (BSI, 2010). PAS 2060 applies to organizations of all types and allows for carbon neutrality across all areas, including buildings, transport, manufacturing, product lines and events.

Conclusion

Following the precautionary principle, climate challenge is such an urgent and serious issue that we should act even in the absence of perfect information. Carbon accounting, like other management tools, may not be an exact science, but its contribution has never been more important. Now that (several) billion tonnes of CO_2 have accumulated in the atmosphere it is time to understand carbon flows and stocks and take action. We must do so if we are to avoid dangerous and irreversible climate disturbances, with unknown consequences. GHG-emission calculations and inventories are a vital first step towards mastering climatic risk. Significant progress has been made since the early 2000s: increasingly rigorous national accounting was developed after the Kyoto Protocol came into force; various technical reports and IPCC assessments were published, the GHG Protocol was developed, ISO standards came out, and tools for raising awareness were developed. With the recent Paris agreement, developing countries are also creating new capacities to calculate their emissions. Despite this progress, methods for calculating emissions vary significantly in techniques and results. The results obtained depend on the application of specific conventions, and there is still a lack of clarity on elements as vital as the radiative forcing of certain physico-chemical phenomena, the definition of boundaries or the level of responsibility for diffuse emissions.

There is no easy or established method for evaluating the cost of present and future damages caused by climate change. Many parameters that are difficult to assess must now be included in evaluation of the long-term policies of states and key private economic actors. There remains uncertainty about the impact of climate change, long-term discount rates, the monetary value of a human life or biodiversity, medium- to long-term adaptation costs etc. However, this imperfect assessment is vital if we want to take into account the true cost of climate change.

These uncertainties about "physico-chemical" assessment of the causes and "monetary" assessment of effects have not stopped the development of policies and measures in order to apply the polluter-pays principle. Climate policies have often taken a traditional approach, using environmental regulations and fiscal instruments like carbon taxes or ecological bonuses, but sometimes they are more innovative, such as the development of carbon markets. Although these new instruments should in theory be effective at achieving a specified reduction target, recent events, such as the carbon-price crash in particular, have shown that in practice design errors or lack of ambition mean that such instruments cannot finance the transition to a new low-carbon economy.

GHGs emissions are caused by manufacturers, transporters, retailers, consumers, public authorities, our actions and interactions. However, our incomplete understanding of these effects prevents us from acting where reductions are easiest or most profitable. Where carbon accounting indicates the sources of emissions and the strategies to follow, this information is sometimes difficult to understand or incompatible with other commendable objectives (such as freedom of movement for people and goods). For this reason it is vital that carbon accounting be freed from the constraints of expert debates, and included in strategic discussions in private companies and in political discourse, so that it can guide the ongoing process of transition.

Many people in leadership roles remain ignorant about CO_2, although it will probably be one of the defining issues of the 21st century. Awareness and understanding of the causes and issues are essential elements in the fight against anthropogenic GHG emissions. Current climate policies and measures were essentially triggered when the scientific community raised alarm, and they take a top-down approach. However, it would be a mistake to think that authorities can develop costly and complex systems of binding regulations without the support of public opinion and various economic actors. The public must understand and support necessary changes in habits and rules, affecting mobility, energy generation systems and consumption patterns.

I hope that carbon accounting education will help the discipline to progress and that the fight against GHG emissions will herald the start of a new economic era in which the search for prosperity can live in harmony with the environment.

Annex
A practical guide to calculating, reducing, offsetting and disclosing your carbon footprint

Arnaud Brohé and Kenneth Hellberg

The benefits of calculating your carbon footprint

It has become imperative that firms address and confront their environmental impact. While demand for sustainable brands has risen as a result of growing concern over the impacts of climate change, increasingly strict governmental regulations throughout the world no longer allow firms the luxury of inaction. While in the past sustainable organizations have benefitted from projecting a green image, this clean image is quickly becoming the status quo.

A carbon footprint allows a firm to quantify its environmental impact, which then allows the creation of actionable goals and the synthesis of useful data. What is not measured cannot be corrected. For example, a firm may find dozens of opportunities for efficiency improvements while performing its carbon footprint analysis. Closing these efficiency gaps helps meet sustainability goals and also saves capital and resources.

For those firms that are required to report their emissions, a carbon footprint helps ensure compliance. If a firm chooses to release the findings of its carbon-footprint analysis to CDP or another agency, the process is streamlined. Firms may also study the reports of other organizations as a means of benchmarking their own findings. Based on the performance of their peers, further evidence for improvements may be uncovered in this analysis.

In addition to competitiveness and profitability, the benefits of a carbon footprint include better access to investment funding. Sustainable and Responsible Investment (SRI) is a growing phenomenon in investing, and as of 2014 more than $6 trillion in assets adhered to its principles in the United States alone (US SIF, 2014). SRI indices were even established. The Principles for Responsible Investment (PRI), a non-profit organization that compels firms to comply with sustainability principles and account for their impacts, now includes more than 1,500 signatories and represents approximately $60 trillion in investments (Principles for Responsible Investment, 2016).

Gathering Data

In order to begin the process of calculating a firm's carbon footprint, it is first necessary to decide the scopes that will be included in this assessment. As is clear in Figure A.1, calculating a firm's carbon footprint becomes increasingly difficult as scopes 2 and 3 are added to the assessment. Also, it is likely that activities that are found under scope 2 or 3, such as business travel and visitor's travel, may be double counted by other firms. For example, if an airline is assessing and offsetting the carbon footprint of a flight through fees on an airline ticket, then it is possible that a firm may still assess the carbon footprint of that flight and offset it through their own efforts. Although these efforts certainly won't hurt the environmental efforts of the firm, the example illustrates how the accountability of emissions becomes more complex as we move away from scope 1 (see Figure A.1).

When contemplating an analysis of its scope 3, a firm needs to consider the scale at which they will calculate the footprint of their products. There are two options: a company-wide footprint, where the emissions are calculated from the top down; and a collection of product footprints, where the emissions are calculated by adding the footprints of all the products a firm offers.

There are strengths and weaknesses to each of these scales, however. The company-wide footprint is faster, more accurate and easily managed, but it may lack the detail needed to identify opportunities for improvement throughout the firm's supply chain or operations. The collection of product footprints allows a firm to analyse its supply chains and operations in great detail, which helps facilitate improvements and innovation throughout the organization. However, these analyses can take a significant amount of time and resources to implement and create complexities when rules are needed to allocate the impact of co-products. In addition, they may not be as impactful for smaller organizations. For example, when a bicycle company that sells one million units a year institutes efficiency changes that lead to a 5% reduction in energy use, the analyses can help save that company hundreds of thousands of dollars. But if a small bicycle company that sells a few hundred bicycles a

Figure A.1   The accountability of emissions becomes more complex as we move away from scope 1.

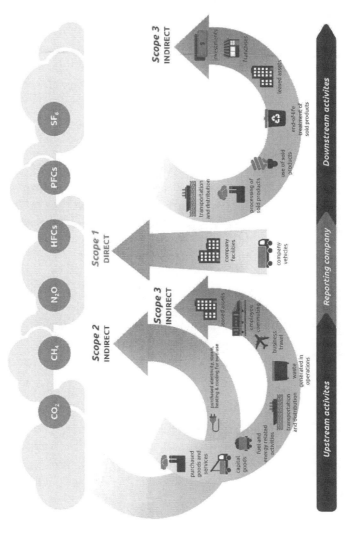

Source: Reproduced with kind permission of WBCSD and the World Resources Institute

year makes the same reductions, it may save less than one thousand dollars. For the smaller company, a company-wide footprint would have been more effective and less costly.

There are other weaknesses to the collection-of-footprints approach. The detail needed to perform life-cycle analyses involves gathering data from suppliers and contractors, which is significantly more difficult than gathering intra-organization data. Further, the accuracy and detail of the data provided by suppliers and contractors may not be accurate unless they have been actively accounting for their energy usage and carbon emissions. These obstacles add a level of uncertainty to the additional costs of this approach.

Therefore, a top-down, company-wide footprint is the simplest and most effective method for building a path toward becoming carbon neutral for organizations who have just begun accounting for their emissions. This top-down, company-wide approach is detailed in the rest of this section.

To begin, indicators must be defined, and monitoring must be established. Depending on a firm's activities, indicators may be either narrow or broad in scope. Indicators may include consumption/generation of energy, chemical/material waste, water usage, paper needs, travel types and distances, etc. To once again use the bicycle manufacturing company as an example, we can quickly identify a few indicators. There is aluminium, steel, plastic and rubber production needed for the bicycles; cardboard is needed to box the bicycles; factories and offices need to be powered, operated and maintained; water is needed for facilities, heating/cooling, manufacturing activities etc.; and there will be chemical and material waste throughout the organization. These are just a few examples of the many sources of carbon emissions in an organization.

Once these indicators are identified and selected to fit the framework of the scopes selected, monitoring methodologies should be

established. Many firms will either appoint a sustainability manager or broaden the responsibilities of another qualified employee.
Firms who are most steadfast in their sustainability goals will take
this even further. Etsy, for example, has established a "sustainability
commission" composed of more than 70 employees, which is led by
a sustainability director (Etsy, 2016). The sustainability manager
must organize the selected indicators and design a plan that moni-
tors their carbon intensity. To use cardboard as an example, a sustainability manager should track overall cardboard purchases and
determine the factors used to determine the carbon intensity of the
cardboard. Fully recycled corrugated cardboard, for example, is
less carbon-intensive than non-recycled corrugated cardboard.
They will have different factors that reflect this.

Data from all indicators should be organized and tracked
separately, but they must also be combined and summarized.
Spreadsheets must be created and updated regularly in a manner
that illustrates the performance of all indicators. Tracking is usually
performed monthly; less frequent tracking may invite problematic
analysis. For example, if a firm tracks its electricity usage quarterly,
then it may miss spikes in usage that might otherwise be mitigated.

Once monitoring practices are established they must be docu-
mented and standardized. This will ensure program compliance
within the organization while maintaining the integrity of the program's data. A training program for all participating employees
should also be implemented to maintain these standards. It is also
a good idea to educate all of a firm's employees on monitoring
programs, as it may help foster a culture of conservation.

When this program is established and data is gathered and organized, a firm can then begin reaping the rewards of its efforts.
Reduction plans can be identified and implemented, offsetting
strategies can be coordinated and pursued, and public-relation
campaigns can be designed and launched.

What data should be gathered?

The data gathered depends on the scopes selected for the carbon footprint. They are detailed below and were categorized by the GHG Protocol (WRI and WBCSD, 2013):

Scope 1

These are emissions directly generated from a firm's sites

- On-site power generation

- Natural-gas consumption

- Refrigerant leakages or other fugitive sources

- Vehicle fuel consumption (performing business activities)

- Manufacturing activities that cause emissions

Scope 2

These are indirect emissions related to energy consumption

- Electricity consumption

- Purchased heat or steam

Scope 3

These are indirect emissions related to all of the other ancillary sources that are inherent in the firm's course of operations. Since its revision in 2013, scope 3 comprises 15 categories for which a specific guidance document is available. We have highlighted in bold the categories that are commonly reported:

　　1. Purchased goods and services

2. Capital goods

3. Fuel- and energy-related activities

4. Upstream transportation and distribution

5. Waste generated in operations

6. Business travel

7. Employee commuting

8. Upstream leased assets

9. Downstream transportation and distribution

10. Processing of sold products

11. Use of sold products

12. End-of-life treatment of sold products

13. Downstream leased assets

14. Franchises

15. Investments

The selected categories should be tracked individually and organized into a dedicated monitoring system. Results should be easily summarized and illustrated in a manner that is conducive to enacting reduction efforts.

Example: Potato Bikes

Let's go back and use the bicycle manufacturer as an example. This hypothetical bicycle manufacturer, Potato Bikes, has three offices and two plants. Potato Bikes produces 10,000 bicycles a year, which are sold in the United States, Canada and Denmark. To produce these bicycles, the manufacturer consumes 125 tonnes of aluminium, 25 tonnes of steel, 5 tonnes of rubber and 4 tonnes of plastic. Each bicycle is wrapped in plastic and protected in a cardboard box with styrofoam. For packaging the manufacturer consumes 50 tonnes of cardboard, 2 tonnes of plastic and 2 tonnes of styrofoam. The three offices are located in New York City, Montreal and Copenhagen. The two production plants are located in Buffalo, New York in the United States and Copenhagen in Denmark. Per month, the facilities consume an average of 70,000 kWh of electricity and an average of 50,000 kWh of natural gas (or 1,706 therms) to power and heat the facilities.

It is easy to look at these numbers and feel a little overwhelmed by how a carbon footprint might be calculated, but a well-organized monitoring program can make this process much more palatable. Let us break down some of the indicators discussed in this example:

TABLE A.1 **Aluminium and steel consumption and carbon footprint in Q1**

Aluminium	Quantity (metric tonnes)	Emissions factor (kg CO_2e/kg)	Total emissions (metric tonnes CO_2e)
January	9	2.0	8
February	10	2.0	20
March	12	2.0	24

Steel	Quantity (metric tonnes)	Emissions factor (kg CO_2e/kg)	Total emissions (metric tonnes CO_2e)
January	2	1.5	3
February	2.5	1.5	3.75
March	2	1.5	3

TABLE A.2   Gas consumption in Q1 (absolute and relative)

Gas consumption in NYC	Square feet	Natural gas use in kWh	kWh/sq ft
January	4,000	5,500	1.375
February	4,000	5,200	1.3
March	4,000	5,000	1.25

Did you notice how we included a table for the square footage of the offices and their energy intensity (kWh/sq ft)? Tracking this data allows a firm to determine the energy intensity of its offices and benchmark the results. This can help detail energy usage throughout the year and analyse the effectiveness of improvement efforts.

Taking this even further, opportunities for improvement can be unveiled by breaking down indicators on a site-by-site basis. See Table A.3 for natural gas:

TABLE A.3   Natural gas consumption and carbon footprint in Q1

Natural gas	US office total use (kWh)	CA office total use (kWh)	DK office total use (kWh)	Plant 1 total use (kWh)	Plant 2 total use (kWh)	Total (kWh)	Total emissions (metric tonnes CO_2e)
January	5,500	6,000	2,500	20,000	26,000	60,000	12.3
February	5,200	6,300	3,500	25,000	15,000	55,000	11.3
March	5,000	5,500	3,500	25,000	18,000	50,000	10.3

By adding more detail to the monitoring protocol, it is easy to see where improvements can be made. Tables A.2 and A.3 clearly indicate that gas usage is steadily decreasing. Should we assume that this is a result of efficiency improvements? Not necessarily. In this case, natural gas is used to heat the offices and the plants. In this hypothetical scenario, January was the coldest month in North

America, and February and March were a little warmer. This reduced the demand for heating in the facilities as the months progressed. However, we see that the Danish office consumes the least natural gas. Let's suppose that Copenhagen also happens to be the home of the biggest office. Climate will certainly impact these results, as Montreal experiences colder winters than either Buffalo or Denmark. But it is clear that there is room for improvement in the other two offices, and that lessons can be learned from the office in Copenhagen. Let's assume the office in Denmark is located in a LEED Gold office building, and that the other two offices are lofts located in reconverted warehouses. Room for improvement is clear, and future efficiency measures can be planned.

Quantifying other scope 1 indicators may prove more difficult, as they are not as simple as looking at purchasing summaries or utility bills. To accurately determine the leakage from air-conditioning systems, it is necessary to know what models they are and what types of refrigerant they use. Older models will use refrigerants that have a much higher GHG potential than newer models, and they also leak more. The vehicles the firm owns can be accounted for by adding up their fuel consumption.

Similar discoveries can be gleaned from scope 2 emissions, which in many cases are just as significant as those found from scope 1 emissions. In the case of Potato Bikes, all facilities purchase electricity from the local utilities. You might think calculating this step might be as simple as adding up all of the kWh of usage and multiplying them against an emissions factor, and you would be mostly correct. It is a little more complex than that, however, as different areas generate their power with different mixes of technology. For example, Copenhagen generates much of its power with wind energy, Montreal generates much of its power with hydro and Buffalo generates a lot of its power with either hydro or nuclear. Buffalo stills uses a significant amount of natural gas and coal (US EPA, 2015); therefore the power consumed in that office may be more carbon-intensive than either Montreal or Copenhagen. To mitigate this disparity, there are different factors for different states, regions and countries depending on the carbon intensity of their electricity grids. For the most accurate results, the local factors must be used.

As Table A.4 shows, it is a fairly simple calculation once the appropriate emissions factor is identified. To find the total scope 2 emissions, we simply calculate the emissions for every site and add the results. These factors may change every year, so it is important to verify that factors are up to date when calculating a firm's annual carbon footprint. For example, suppose Denmark installs a colossal offshore wind farm that it allows its electricity grid to consume 95% renewable energy. The emissions factor will drop from the previous year. On the other hand, if a new coal plant is completed, the emissions factor in Denmark will most likely spike.

TABLE A.4   Electricity consumption and carbon footprint in Q1

Electricity (US Office)	Total kWh	Emissions factor (US EPA, 2015) ($kgCO_2$/kWh)	Total emissions (metric tonnes CO_2e)
January	5,500	0.315	1.73
February	5,400	0.315	1.70
March	5,300	0.315	1.67

Scope 3 emissions are much more difficult to track than scope 1 and 2 emissions, as they are not generated directly by the firm and therefore cannot be calculated with simple accounting practices. We will analyse each of the major indicators a little more closely for clarity:

Employee Commutes

This may be calculated by polling employees about their modes of commute. Depending on how far they live from the job and how they commute, a lot of factors will determine this final calculation. The data from the poll can be summarized and organized in a way that will simplify the calculation. For example, there are three variables at play in this indicator: mode of commute, distance of commute and frequency of commute. All employees should be grouped into their modes of commute. Next, employees should multiply their daily commute distances by the number of days per month they commute to work. The total distance should be averaged by their respective modes and then multiplied by an emissions factor in order to calculate each mode's emissions. These emissions can be added for the total figure.

Waste treatment

Decomposition of solid waste in landfills and wastewater treatment contribute a significant proportion of methane emissions EIA, 2016) and therefore form a component of a firm's emissions profile. Waste treatment activities may include:

- Disposal to landfill

- Recycling

- Incineration

- Composting

- Wastewater treatment

It is important to distinguish between facilities owned or controlled by a third party (scope 3 emissions) or waste-treatment facilities owned or controlled by the reporting company (scope 1 or 2 emissions). Third-party waste-treatment companies can potentially supply their waste-specific emissions data to companies. However, for waste-treatment facilities owned or controlled by the reporting company, the following data must be collected:

- Waste produced (e.g., tonne or cubic meter)

- Type of waste generated (e.g., plastic, water, sludge, food, paper or rubber)

- Waste treatment method applied (e.g., waste to landfill, incinerated, recycled)

Emissions are calculated using factors specific to waste-treatment emission or blended emissions factors by industry (e.g., municipal or industrial waste-emission factors) depending on the granularity of the data collected.

Freight

Freight can be measured by separating the emission figures into separate calculations based on the type of freight. For example, Potato Bikes uses ground truck delivery to deliver its bikes to the central delivery facility in the United States, but it uses a train to deliver its bikes to the central delivery facility in Denmark. From those facilities, bicycle shipments are parcelled out and delivered to distributors. The distance and weight of the shipments via their respective modes of transportation should be organized, and a carbon calculation can be made by researching and applying the most accurate factors.

Extraction and production of all purchased resources and materials

As detailed in the tables in this appendix with figures, this carbon calculation can be made by multiplying total resources and materials purchased by their respective factors. Since every resource or material has a different emissions factor, the carbon calculation for each type of resource or material should be performed separately.

These figures can be compared with waste figures on a regular basis to ensure that waste-reduction efforts are effective and that new waste is not finding its way into the firm's operations.

Transmission and distribution losses

According to the World Bank (2014), the world loses about 8% of the electricity in transmission and distribution losses. This figure is different from country to country, however. The United States loses about 8%, Canada loses about 9% and Denmark loses only about 4%. These losses can be accounted for by simply adding the lost percentage to their respective electricity-related emissions results.

Paper-related consumption

Many overlook the potential for improvement in paper consumption. According to the US Forest Service (US Department of Agriculture Forest Service 2014), per capita consumption of non-fuel wood and paper products is more than 60 cubic feet per year (or 1.7 cubic metres). Although paperless services and processes reduce this number dramatically every year, there is still significant reduction potential.

For this carbon calculation, the types of paper a firm uses (weight, material) should be tracked separately, and the pounds of each type of paper consumed should be accounted for. These totals should be multiplied by their respective emissions factors.

Where do I find the proper emissions factors?

The most appropriate emission factors to use will depend on the goal of the organization. For comparability among peers, it will be important to use industry standard or local regulatory frameworks if available. Firms with operations in multiple jurisdictions and countries should calculate emissions based on the most up-to-date research produced by a reputable source such as country averages published by the International Energy Agency. Alternatively there are country- or region-specific sources such as the US Energy Information Association (EIA), the European Environment Agency, the National Greenhouse and the Australian Government Department of Environment National Greenhouse Accounts. The Swiss database Ecoinvent is probably the best source of information for secondary emission factors.

How should I interpret this data?

A carbon-footprint analysis allows an organization to analyse, compare and prioritize emissions sources. Just as there are many opportunities to generate emissions, there are many opportunities to curb them. A firm might find that it can reduce its energy consumption, or perhaps it may attempt to reduce waste, which will help reduce both consumption and waste-treatment emissions. Steps should be taken to pursue the most impactful actions first.

Comparing the emissions sources can aid in prioritization, but it can be difficult to discover if performance in any given metric is satisfactory. Benchmarking is a tool to overcome this. Many cities have begun requiring that firms with large commercial spaces report their energy use in an effort to benchmark their efficiency. These data can be used to compare a firm's energy use with that of its peers.

Once a firm identifies areas of weakness, it can prioritize its measures of action. Some of these measures may prove more costly than others. In all cases, a simple cost–benefit analysis should be performed to judge the feasibility of each. For example, purchasing some low-flow aerators for the faucets in a facility is an inexpensive way to reduce both water and water-heating–related energy consumption. Just a few dollars can make a noticeable impact. On the other hand, replacing an old, inefficient boiler can be extremely costly. While the boiler may pay for itself after a decade, if a firm is running lean then this step may not be feasible.

It is important to remember that while a carbon footprint is a direct measure of a firm's impact to the climate system, it is also an indirect measure of a firm's overall efficiency. Taking action to reduce a firm's carbon footprint results in lower overhead costs, reduced materials expenditures and improved profitability. Couple these improvements with an effective communications plan, and a firm can garner positive press and recognition.

How should a carbon-footprint report be designed?

A carbon-footprint report should be designed in a manner that allows it to be included, if required, into any corporate statement. To accomplish this, the calculations should conform to the ISO 14064 and ISO 14069 standards as well as the GHG protocol. This is more technical than simply calculating the carbon footprint, and for firms who are reporting for the first time, hiring an experienced consultancy will be more or less a necessity.

To meet these standards, a carbon-footprint report will:

- rank carbon emissions by source

- present relevant charts and key performance indicators

- enable a firm to clearly identify sources of carbon emissions

- recommend actions that can be taken to reduce carbon emissions

- consider and assess the effect such actions will have on reducing carbon emissions

Which framework should I use?

The GHG Protocol is the most widely used international standard for quantifying GHG emissions and forms the basis for almost every GHG standard in the world. It was developed in partnership by the World Resources Institute (WRI) and the World Business Council for Sustainable Development (WBCSD). Any listed companies must consider reporting their carbon footprint to the Carbon Disclosure Project (CDP). Eighty-six percent of Fortune 500 companies reported to the CDP using the GHG Protocol in 2014.

External third-party verification or assurance is recommended for best-practice reporting. Look for auditors who can provide assurance against the ISO14064 standard.

Should I include other environmental impacts from the start?

Many of the processes tracked for a carbon footprint overlap with energy-, waste- or water-intensive processes, making it easy for companies to streamline procedures to collect data on multiple environmental impacts together. For example, once a company starts tracking waste to landfill to calculate waste-treatment emissions, the data can also be used to report on quantities of waste produced.

A full Life-Cycle Analysis (LCA) includes indicators such as eco-toxicity, smog and heavy metals; these are difficult for company executives, customers and third parties to understand. This level of detail may be daunting or dissuade organizations from pursuing environmental initiatives. We would not recommend starting your sustainability journey with an LCA unless you need to eventually develop an Environmental Product Declaration.

If you have already completed a materiality assessment, it would be prudent to report on all material environmental impacts. For example, if air pollution or water use appear to be more important than climate-change impacts it might be worthwhile examining these aspects over the life-cycle of your products. Materiality assessments are a crucial pre-cursor to writing and publishing a sustainability report. Often combined with stakeholder engagement to determine what matters most to a company's stakeholders, materiality assessments inform a company on what non-financial data it should be reporting on and why. If you decide to release a sustainability report, the Global Reporting Initiative (GRI) framework is the gold standard for sustainability reporting. GRI provides a wealth of reporting guidelines, instructions and templates to help guide your report structure and content.

Should I hire a consultant?

A firm has two broad choices when it comes to performing a carbon footprint: delegate the responsibility to competent employees, or hire a consulting agency. There are benefits to both approaches, and in some ways they need not be exclusive.

As mentioned earlier, any large firm that is serious about its sustainability goals should at least have a sustainability manager. Smaller firms may not have the operational scale that warrants this position, but they can still educate their employees about sustainable practices and enact sustainability programs. A sustainability manager will cost his or her salary, but educating employees may require hiring a consultant or a third-party firm. Educating employees will also cost time and productivity, as they will be performing new tasks with which they are not yet comfortable, possibly hindering their primary roles. Either way, standardized, accurate reporting results are necessary for the program to be effective and, if applicable, within compliance.

Hiring a consulting agency can prove beneficial, but it is necessary to choose wisely. A consulting agency should have proven experience in performing carbon footprints and reports for a diverse array of firms, as every firm is unique and will require analytical agility. The agency should also be able to provide solid references and detailed proposals. A consulting agency should be a necessity for organizations that have never performed a carbon-footprint analysis and have yet to install a sustainability plan. If the firm has hired a sustainability manager, the consultant can help ensure that the initial footprint is accurate and pertinent to the firm and its goals. The consultant can also help write the sustainability plan, create any educational materials, design marketing and outreach materials, and connect the firm to organizations that will help it achieve sustainability goals.

When considering costs, it comes down to choosing between time and human resources and a capital investment. If a firm is unwilling to invest in hiring or training sustainability professionals, then it must hire a consultant to perform these tasks. In either case, an ongoing relationship is necessary to ensure the accomplishment of sustainability goals. Data that a consultant gathers and interprets must also be clearly understood by C-level members of the firm; otherwise opportunities to improve efficiency and competitiveness may be overlooked.

How do I convince my boss?

Firms are loath to part with money unless it helps their bottom line. While performing a carbon-footprint analysis can help uncover room for improvement, these results are neither immediate nor guaranteed. But it may not take much convincing – 93% of CEOs consider their firms' reaction to sustainability issues critical to their firms' future success (Project Management Institute, 2011). Fortunately, there is observed data that can help convince even the most ardent skeptics.

According to Nielsen's 2015 Sustainability Imperative, 66% of consumers reported that they were willing to spend more for brands that were sustainable. This statistic was clearly illustrated in market performance, as sustainable brands significantly outperformed brands that failed to demonstrate a commitment to sustainability (Nielsen, 2015). Furthermore, 45% of consumers reported that a key purchasing driver for them was a firm's environmentally friendly image (Nielsen, 2015).

These numbers are only increasing, especially as younger consumers are generally more likely to demand sustainability in their products. In fact, 51% of millennials check packaging for sustainability claims before purchasing a product (Nielsen, 2015). More and more, consumers considering products and services may reject brands that neglect their sustainability commitments.

Businesses that fail to address sustainability may also find themselves unable to attract and attain top-level talent. Of students polled, 92% said they would seek employment with an environmentally friendly company, and 75% see a firm's CSR commitment as a factor in selecting an employer (Willard, 2012). Firms that neglect sustainability also experience increased employee attrition rates. In 2007, 57% of employees said that a firm's CSR commitment was a factor in staying with the firm (Willard, 2012). Just as consumers have become more attracted to sustainable brands, it is reasonable to assume that employees will as well.

After looking at this data, the question should no longer be about whether or not a boss can be convinced, but whether or not that employer can afford to ignore sustainability. The carbon footprint is the first step in measuring a firm's impact, which can then aid in formulating a strategy to reduce that footprint. Consumers will notice, and employees will notice. If firms ignore sustainability, they risk consumers and employees ignoring them.

How to engage my colleagues?

Whether you serve directly as an example or facilitate greater awareness of carbon issues, there are a number of ways to engage your colleagues. It is understandable that many people want to avoid irritating their bosses or alienating their peers. Luckily, this is easily avoidable.

First, you can raise awareness to climate change and carbon impacts by caring for them in your actions. When people eat low-impact foods, ride a bicycle to work, avoid wasting paper or other consumables, use a refillable water bottle or bring to work vegetables that they grew in their own garden, these behaviors are noticed by others. Turning lights off and shutting down computers when they are not in use and separating your waste into proper recycling receptacles are both simple actions that influence co-workers. For example, it is highly unlikely that a co-worker will throw paper in a garbage bin after he or she has seen you recycle yours. By taking care to live as sustainably as possible, you in turn passively influence those around you.

Second, you can raise awareness by participating or engaging in events related to sustainability. For example, Earth Day is 22 April. You might e-mail your peers with a happy Earth Day message or even hold a "birthday party" for the Earth. Cities will have a number of events on or around Earth Day, and you can help bring your peers' attention to these events by inviting them. In addition, some cities have events that are centred on sustainability. For example, New York City has a Climate Week when there are dozens of events to bring awareness to this issue. In addition, near the end of every year the UNFCCC Conference of Parties (COP) holds its meeting. Following the COP events and sharing the outcomes with peers is a great way to bring recognition to how important sustainability is to the international community.

By setting an example and engaging peers, you are likely to attract interest in your behaviour, and these are opportunities to inspire climate action.

How to reduce emissions in buildings

According to the EIA's 2016 International Energy Outlook, residential and commercial buildings consume over 20% of the world's delivered energy, and that share is poised to grow. The commercial building sector is projected to lead growth in energy demand (EIA, 2016). Electricity and natural gas deliver more than 80% of the energy for this sector, and that number is also expected to grow (EIA, 2016). In this way, efficiency measures in buildings will play an important role in mitigating carbon emissions by stemming the development of a growing source.

Firms will increasingly be rewarded for reducing the energy use in their buildings with lower energy bills, fewer equipment expenses, and bolstered sustainability efforts. It is estimated that energy accounts for 19% of the average office building's expenditures (National Grid, 2002). This means for every 5% reduction in energy use a firm can realize a 1% drop in facility costs. For larger facilities, this can represent tens of thousands of dollars in annual savings. A 2009 US Department of Energy study found that a "reasonable range" of potential energy savings in commercial buildings is anywhere from 10% to 20%.

There are three primary ways a firm can reduce the footprint of the space that it occupies: economize its space needs; increase the efficiency of its space; and generate the energy consumed by its space. All three should be considered, but the first two are applicable to almost every firm.

Economizing occupied space can be accomplished in a number of different ways, which can help reduce fuel costs, maintenance costs and, if applicable, leasing costs. If it involves selling or leasing unused space, it may even become a source of revenue. While a firm thinks most of the space it occupies is useful and productive, it may not necessarily be the case. A closer analysis could reveal that a collection of vacant spaces in its building, if combined, will allow the firm to divest or sublet costly real estate. In addition, firms can consider turning to the internet. Allowing employees to telecommute will not only reduce the space a firm needs, but it will also reduce its scope 3 emissions. Teleconferences and videoconferences allow a firm to avoid business travel and the need for a conference room. When these trimming measures are taken to the extreme, a firm may find that it needs very little space, especially if the scope of its operations do not involve manufacturing. In response to a growing demand for transitory space, shared office spaces have spread in many cities, allowing firms to access offices and boardrooms when needed.

Efficiency must be pursued alongside the mitigation of unneeded space. There are two ways to increase energy efficiency in a building: behavioural change and technology or equipment upgrades.

Behavioural change requires analysis and education, and employees should be trained to perform tasks efficiently and in a manner that minimizes energy use. The first step is to clearly communicate the firm's sustainability goals, and then the next step should be to teach common, no-cost habits that can save energy. These habits will change from firm to firm, depending on the facilities they operate and the duties of its workforce. This is the least expensive method of increasing efficiency in a facility.

Each role in an organization has its own broad, unique set of responsibilities. In the same way that these responsibilities are examined for performance assessments, they should also be analysed for efficiency opportunities. For example, if a technician is responsible for visiting different sites throughout the week, then these trips should be arranged in a way that can both satisfy performance needs and reduce travel distance to a minimum. By integrating an efficiency-potential analysis with a job performance analysis, firms can decrease energy use as they increase worker productivity.

Technology or equipment upgrades may prove more costly upfront, but they often pay for themselves and establish energy reductions for many years. These upgrades can be as simple as screwing low-flow aerators onto faucets, or they may be as complex as installing automation and control equipment on consumptive components of mechanical systems. Often, technology can help overcome behavioural limitations, such as when light sensors and equipment timers are programmed to shut things off in cases where people might otherwise forget.

The least expensive measures should always be taken first, as they will involve fewer barriers. More expensive measures may require budgetary planning or even financing. To determine the payback period and feasibility of these upgrades, a detailed energy audit should be performed and energy performance should be benchmarked against similar facilities. This will help identify the most effective and practical strategy for increasing technology and equipment efficiency.

Lastly, a firm should consider generating its own technology with renewables. The potential for this is often dictated by the buildings or land that it owns or leases. Not all firms have the luxury of owning their facilities, and many more find themselves leasing space in large office buildings. However, if a firm owns property that is suitable for solar PV, it is a step that should be considered. For firms that generate their own power for manufacturing operations, combined heat and power can be an extremely beneficial technology. Geothermal heat pumps can help drastically reduce energy needs for heating and cooling, which typically account for most energy demand. Solar water heaters can also help reduce energy needs.

If a firm leases its space and finds that there are significant opportunities for efficiency, but it is unable to secure approval for the installation of renewable energy or more efficient technology or equipment, then it can consider either changing the terms of a renewing lease or searching for a new location. "Green" leases have been developed, and they allow a firm to integrate its environmental efforts into its leased space.

If a firm decides a new location is more feasible, then this allows for a reinvigorated consideration of efficiency. For example, if a firm finds that most of its employees live in a town 30 miles from the building, then it may consider moving closer to its employees. Another possibility is to move closer a train station, which then allows employees to take advantage of mass transportation.

How to reduce emissions in personal transport?

There are a number of ways to reduce your emissions in your personal transportation, which will in effect reduce your firm's scope 3 emissions. Steps to reduce your transportation emissions range from high-cost to no-cost options; however, every step helps reduce travel-related fuel expenses.

The most impactful choice in transportation is how you choose to travel. For example, riding a bicycle to work will erase all of your commute-related emissions. But for many, this is not a practical option. Mass transportation options such as the bus or the train tend to be less carbon intensive as well, especially if a city uses hybrid or alternative fuel technologies to power their buses or they generate their electricity with clean generation technologies to power their trains. Carpooling is another fantastic option, and it allows many people to access High Occupancy Vehicle (HOV) lanes found in many cities.

If a person must rely on a personal vehicle for commutes, then the next most impactful choice for reducing emissions is vehicle type. Choosing an electric vehicle is the lowest-emission option; however the high price of the vehicles and the lack of charging infrastructure may rule this option out for many people. Hybrid and plug-in hybrid vehicles are also excellent options. Although they tend to be more expensive than their gasoline-only counterparts, the fuel savings over the life of the vehicle can be significant. Also, because they have been mass-produced for a long time, there is a large market of used hybrid vehicles. Smaller vehicles are also more efficient than their larger counterparts. It is also prudent to research the fuel economy of vehicles.

Driving habits and vehicle maintenance can also play an important role in fuel economy. Aggressive driving is fuel-intensive, and accelerating or braking more rapidly than necessary is wasteful. Smooth, slow acceleration and cautious, prudent braking techniques can significantly improve fuel economy. Also, the speed a person drives can have a big impact on fuel economy. This is because wind drag on a vehicle increases as speed increases, which makes the engine work harder and harder just to maintain the vehicle's higher speed. Checking tire pressure can also help increase fuel economy. Maintaining full tire pressure reduces rolling resistance. In addition, keeping an engine properly maintained and tuned can also increase fuel economy. The US Department of Energy (2016) estimates that addressing maintenance problems can have huge impacts. For example, fixing a corrupted oxygen sensor can improve fuel economy by up to 40%.

How to reduce emissions in logistics?

Logistics-related emissions account for an increasingly large share of commercial-industrial emissions. In fact, US freight emissions may increase by 40% above 2014 levels unless firms begin making emissions-reductions in their logistics a priority (Mathers *et al.*, 2014). Reducing emissions may often equate to reduced transportation expenses and more efficient supply chain operations.

The Environmental Defense Fund suggests five strategies to reduce emissions (Mathers *et al.*, 2014):

- **Maximize utilized space in all freight vehicles and shipped containers.** It is estimated that 24% of freight vehicles are travelling empty, while the remaining vehicles are filled to 57% capacity (Jesus Saenz 2012). There is a significant opportunity for improvement in this area.

- Choose the least carbon-intensive mode of transportation. Ocean transport is more efficient than rail transport, while rail transport is more efficient than truck transport.

- Collaborate with suppliers, vendors and competitors. There can be opportunities to capitalize on synergies that may have been overlooked.

- Review and optimize the logistics network. A firm's logistics network should be inspected for improvement opportunities regularly. Things change, and new developments in how a firm operates can change logistics needs dramatically. Failing to optimize logistics networks to keep up with changes in operations can lead to significant inefficiencies.

- Work with logistics providers to inspire efficiency. If a firm is paying for third-party transporters, then it likely has choices for what transporters it will use. Before locking in agreements, it might be a good idea to compare transporters' vehicles and fuel choices. If a transporter is using dated tractor-trailers that burn diesel, then they will emit more emissions than a transporter that uses newer trucks that run on natural gas. The further a firm must transport its product, the more significant these differences can become.

How to educate your employees and engage your suppliers and customers?

A firm's sustainability measures can inspire reductions beyond its operations by educating its employees and engaging its suppliers or customers. Strategies to reduce energy use and waste in the workplace can often translate to the home, and employees are likely to take home some of the practices they learn. For example, if a firm installs low-flow faucets in all of its sinks and tells the employees that this was done to reduce water use and water-heating-related energy expenses, employees discover an opportunity to reduce their own utility bills. Educating employees can be as simple as sending out an e-mail or asking supervisors to discuss these actions at their weekly meetings. More comprehensive sustainability programs will make sustainability a key part of all communications.

How do I set a reduction target?

There are a variety of methods for setting carbon-reduction targets depending on a company's industry or circumstance. For example, certain methods are more appropriate for companies growing rapidly and others for companies with operations in multiple countries. Regardless of the methodology chosen, best-practice climate-reduction targets are in line with science-based targets. One example of this is the sectoral decarbonization approach which was developed by the CDP, WRI and WWF in response to the scientific community's call to limit warming to a 2°C temperature increase above pre-industrial levels. This methodology provides comprehensive guidance for companies to set targets and emissions reductions for a 2°C pathway by sector and takes a least-cost mitigation approach.

Carbon-reduction targets should be realistic within the limits of your available budget and your timeline for reducing emissions. Ideally targets should be tiered for both the short term (five years) and the longer term (ten years). There is no one-size-fits-all approach for setting carbon reduction targets and a company will need to tailor the approach that best suits its needs.

Let's compare two approaches, a revenue-based target or an absolute emissions target. A revenue based target such as $tCO_2e/\$$ is an intensity-based target which reflects the carbon intensity of each dollar generated. An absolute emissions target reduces total emissions, for example a 20% reduction in total emissions across all operations by 2025. The right approach for a company will demand on its growth projections, industry, available investment and technology.

Should I set an internal price on carbon?

The private sector has increasingly taken the charge on incorporating the externalities of climate-change risks into their investment and planning decisions by setting an internal price on carbon. Companies are now realizing the very real risk that climate change poses to their operations either in response to or anticipation of climate-pricing legislation.

By setting a price on carbon, a company can protect its supply chain or manufacturing facilities from increased energy prices or price fluctuations as a result of extreme weather activities. There are three major approaches to price carbon: internal carbon taxes, fee or trading systems; shadow pricing; or direct pricing. Carbon pricing by companies is most often in the form of a shadow price on carbon – that, is a price that can be added to future investments and costs to hedge against future policy decisions to price carbon.

What price you decide to set on carbon will depend largely on the specific conditions your company is operating in, as well as industry or government price projections. For the greatest impact, the carbon price should be high enough to materially affect investment decisions to reduce emissions and ideally they should increase over time. In 2016 the United Nations Global Compact called on companies to set an internal price on carbon at a minimum of $100 per metric tonne over time.

How can I become carbon neutral?

Once you have set a carbon reduction target and set an internal price on carbon, you should start to see a decrease in your carbon emissions. As you work towards being carbon neutral, the final step is to offset your remaining emissions. Although it is possible to develop your own offset program, it is usually more cost effective and efficient to purchase offsets from a project developer or carbon-offset retailer. These can come in the form of carbon credits or Renewable Energy Credits (RECs). Carbon offsets are associated with either removing existing CO_2 from the atmosphere via sequestration from tree planting or the avoidance of future CO_2 emissions from avoided deforestation, energy efficiency or renewable energy projects. When selecting a carbon-offset program it is important to look for the following aspects:

- Real – the project exists as it has been described or sold

- Verified – a third party should have provided a guarantee or assurance that the project meets a transparent set of criteria

- Enforced – there is a mechanism for penalizing the project proponent if the criteria has not been met

- Permanent – the offset should have a guarantee that it will exist into the very long-term future; for example that trees planted won't be removed in six months

- Additional – the carbon-offset action was not going to go ahead without the carbon offset project

There are exciting opportunities for social co-benefits with some carbon-offset projects creating positive social impacts as well as climate change benefits. For instance, CO_2logic supports a unique carbon-offset project in Benin where 90% of the population relies on wood and charcoal to cook their daily meals, thus contributing to forest degradation. The project facilitates access to improved stove technologies which reduces wood and charcoal consumption by up to 50%, saves US$100 per year per family and reduces household air pollution which frequently cause respiratory illnesses such pneumonia.

How do I release and communicate our efforts?

The final and important piece is communicating your progress against carbon reduction targets over time. Clear and transparent communication on sustainability efforts is a powerful way to advocate the importance of climate-change action. Reporting on climate change initiatives is also an important opportunity for companies to respond to pressure from investors and stakeholders on their sustainability credentials.

How a company chooses to disclose this information is at the discretion of the company executives and will depend on a company's broader marketing strategy. Corporate responsibility reports are an important touchstone for a company's current and future employees and provide stakeholders with the knowledge that risks and opportunities are being appropriately managed. CDP and GRI are the two most popular initiatives for carbon disclosure.

However in some jurisdictions, disclosure is not an option. For instance companies under the EU ETS must release their direct emissions on a yearly basis and the data are audited by an accredited third-party verifier and published by public agencies.

Glossary

Additionality: A project to reduce GHG emissions is additional if it cannot be carried out without extra financing provided by the sale of emission-reduction credits. In practice, project developers generally have to demonstrate that there are financial, cultural, technological or institutional barriers and that these barriers can be overcome through the CDM.

Allocation: The quantity of GHG emission rights attributed in a regulated carbon market with emissions caps (the cap-and-trade system). The initial allocation of quotas is limited on the basis of a reductions objective. These can be allocated free of charge (e.g., on the basis of historical emissions or environmental performance) or via a bidding process.

Allowance (or quota): An accounting unit allocated to an entity within the framework of a cap-and-trade system, representing one tonne of greenhouse gas.

Annex I/Annex B: The list of countries in Annex I of the UNFCCC was established by the 1992 climate convention, consisting of 38 developed countries and countries in transition to a market economy. It is virtually identical to the list of Annex B countries in the Kyoto Protocol of 1997, which states specific commitments to which industrialized countries should conform. The only differences are that Annex B includes Croatia, Liechtenstein, Monaco and Slovenia, but leaves out Belarus and Turkey.

Bilan Carbone (carbon balance): The Bilan Carbone and its database of emission factors provide a comprehensive methodological basis. Since late 2011, the Association Bilan Carbone (Carbon Balance Association) has managed licenses for the tool and the Institut de Formation Carbone (Carbon Training Institute) has organized training in the method.

Carbon credit: A carbon credit is a unit representing an emission right, granted during implementation of project mechanisms such as CDM or joint implementation. Carbon credits are also used for voluntary offsetting.

Carbon dioxide: Also known as carbonic acid gas, it is the most significant gas in the anthropogenic greenhouse effect. Used as a measure of GHG emission by means of the common unit, the tonne of CO_2-equivalent.

Cap and trade: A mechanism for limiting emissions in which emission rights are defined and capped, and can then be traded. If a participant does not have enough allowances to match its emissions at the end of the compliance period, the participant must buy additional allowances on the market to make up the shortfall. If the opposite is the case, the participant can sell allowances. This system is sometimes linked to a project mechanism that generates carbon credits that are fungible with quotas if certain conditions are met.

Carbon footprint: A carbon footprint is defined as a measure of total GHG emissions caused directly or indirectly by a person, an organization, a territory, a product or a service.

Carbon leakage: Carbon leakage occurs when a company delocalizes to a country with climate regulations that are more lax or non-existent. This type of situation reduces the emissions of environmentally progressive countries without any environmental benefit at the global level.

CDP: The CDP is an independent non-profit organization that collects primary information from companies about their GHG emissions, carbon-reduction strategies and other environmental issues (water use, deforestation, supply chain). Founded in 2001 by a consortium of institutional investors, the CDP requests information from large multinational companies on a yearly basis.

Clean development mechanism: This mechanism was set up by the Kyoto Protocol. It enables industrialized countries (Annex B countries) to buy certified emission reductions (CERs) generated by projects carried out in developing countries.

Covenant of Mayors: The Covenant of Mayors is a movement that brings together local and regional authorities on a voluntary basis to improve energy efficiency and increase the use of renewable energy in their territories.

Environmental disclosure: The action of making information about the environmental impacts of a product available to consumers, so they can compare products in the same category.

Environmental impact: Any negative or positive change to the environment resulting entirely or partly from the activities, products and services of an organization.

Environmental product declaration: an EPD is a comprehensive, internationally harmonized report that documents the ways in which a product, throughout its life-cycle, affects the environment.

Emission factor: An emission factor shows the quantity of GHGs emitted in relation to the activity responsible, in CO_2-equivalent per accounting unit of the activity (e.g., kg CO2-equivalent/kWh, kg CO_2-equivalent/tonne of steel, kg CO_2-equivalent/cow year etc.).

EU ETS: European Union Emission Trading System. A system for capping industrial emissions, which enables trading of CO_2 allowances among more than 10,000 industrial facilities and airlines subject to emission limits.

Flexibility mechanism: The Kyoto Protocol created three flexibility mechanisms, which were designed to decrease the cost of reducing emissions in line with the ceilings set for states and facilitate implementation. The specific mechanisms are the Emissions Trading Mechanism, the Clean Development Mechanism (CDM) and joint implementation.

Forward: Refers to a forward contract negotiated by mutual agreement in a non-standardized way (in comparison to "futures", which are negotiated in an organized market).

Functional unit (FU): This is the quantified performance of a product system to be used as a reference unit in a LCA. The FU has a time dimension: to ensure the function F during the expected service life of the product. The FU includes the reference flow (in kilograms of product), packaging, fitting accessories, and the scrap rate during installation and maintenance.

Futures: Futures are forward contracts with standardized terms for amounts and maturity dates.

GHG Protocol: The *GHG Protocol* was published in 1998 by the World Resources Institute (WRI) and the World Business Council for Sustainable Development (WBCSD) working in partnership. This multilateral partnership of companies, NGOs and governments proposed the first accounting framework for company GHG-emissions standards, programmes and inventories, and still serves as a benchmark.

Global Covenant of Mayors: The newly created Global Covenant of Mayors for Climate and Energy unites more than 7,100 cities in 199 countries across six continents in the shared goal of fighting climate change through coordinated local climate action.

Global warming potential: The global warming potential of a GHG represents the radiative forcing that a given quantity of gas returns to the earth, accumulated over a period generally set at 100 years. This value is compared to that of CO_2.

Greenhouse gas (GHG): These are natural and anthropogenic gases in the atmosphere that absorb and re-emit infrared radiation. The GHGs regulated by the Kyoto Protocol are carbon dioxide (CO_2), methane (CH_4), nitrous oxide (N_2O), hydrofluorocarbons (HFC), perfluorocarbons (PFC) and sulphur hexafluoride (SF_6). Accumulation of GHGs in the atmosphere is one of the factors affecting the global warming mechanism. Water vapour, ozone and CFCs are other GHGs not covered by the Kyoto Protocol.

Greenwashing: A marketing or public-relations exercise carried out by organizations that want to project an unrealistically positive image of their environmental management.

The Grenelle environment forum: A set of political meetings organized in France in late 2007, with the aim of making long-term decisions on the environment and sustainable development, particularly to restore biodiversity, reduce GHG emissions and improve energy efficiency. Propositions were drafted in several working groups during the first phase. Two laws eventually materialized from this consultation process: the Grenelle Act 1 (2009) and the Grenelle Act 2 (2010).

IPCC: The Intergovernmental Panel on Climate Change is responsible for organizing a summary of scientific work on climate change.

ISO: The International Organization for Standardization is an international standardization body comprised of representatives of national standardization organizations from 162 countries.

ISO 14000: The ISO 14000 series of standards was developed to help companies continuously evaluate and control the environmental impact of their activities, products and services. These standards do not impose absolute requirements for environmental performance, but rather a management commitment to follow legislation and the principle of continuous improvement.

Joint implementation mechanism: Set up by the Kyoto Protocol, it enables industrialized countries (Annex I countries) to buy emission-reduction units (ERUs) generated by projects carried out in other industrialized countries in Annex I.

Kyoto Protocol: When industrialized countries signed this protocol in Kyoto (Japan) in 1997 they made a commitment to reduce their emissions of six main GHGs by an average of 5% between 2008 and 2012, compared to 1990, which was chosen as a reference year.

Life-cycle analysis: LCA is a methodology for compiling and evaluating inputs and outputs of energy and materials, in addition to environmental impacts of a product system throughout its life-cycle, from extraction of raw materials to end-of-life waste treatment. LCAs enable the identification of improvements that can be made during the different stages of the life-cycle.

Life-cycle inventory: The life-cycle inventory (LCI) is a phase of life-cycle analysis (LCA), which involves compiling and quantifying incoming and outgoing flows.

Marginal abatement cost: The marginal abatement cost (MAC) represents the cost of reducing the last tonne of CO_2 required to achieve a fixed emission-reduction objective (when using the cost-efficiency analysis perspective). This instrument makes it easier to compare the cost of pollution reduction in various sectors and thus to evaluate the overall economic cost of a given policy. The MAC is generally presented in the form of a curve illustrating how it increases for each unit of emission reduction.

Marrakesh Accords: Accords negotiated during the seventh Conference of the Parties (2001), which set the principal rules for international exchange of emissions rights and practical aspects of the other two flexibility mechanisms.

Project mechanism: Project mechanisms enable investors in low-carbon projects to generate emission credits. The Clean Development Mechanism and joint implementation are the two most commonly used project mechanisms.

PAS 2050: Published by BSI in 2008, PAS 2050 was the first standardized methodology for calculating the carbon footprint of products or services. This methodology, based on the concept of LCA, enables the conversion of GHG emission flows into CO_2-equivalent for activities in the industrial or service sectors, by product or service type.

PAS 2060: Published by BSI in 2010, the PAS 2060 specification describes ways to demonstrate carbon neutrality and defines three necessary stages (calculate, reduce and offset).

Polluter-pays principle: The legal principle that the polluter should take on the "expenses of carrying out preventive measures decided by public authorities to ensure that the environment is in an acceptable state" (OECD 1972). It is one of the basic principles underlying environmental policy in developed countries. It is also the origin of pollution costs internalization by those responsible for pollution, through regulatory instruments (standards, bans, permits, zoning, quotas, restrictions on use and other direct regulations), economic instruments (charges, subsidies, deposit-return systems, market creation, compliance incentives), or fiscal instruments.

Primary data: quantitative data from direct measurement or from a calculation based on direct measurements of an activity or a process. Primary data shows the nature and specific effectiveness of an activity or a process, providing an indication of its specific environmental impact.

Programme of activities (PoA): a programme that makes it possible to add an unlimited number of small GHG emission-reduction projects by actors, sectors and regions, within the framework of the CDM.

REDD/REDD+: Carbon-finance mechanism linked to the fight against deforestation and promotion of reforestation. REDD+ goes beyond the fight against deforestation and promotion of reforestation – it also encourages sustainable management practices and improvement of forest-carbon stocks.

Registry: Accounting system for emission allowances from allocation to surrender. The register is the large set of accounts for the emission-trading system (allowances and credits).

Shadow price of carbon: The shadow price of carbon is a fictional price determined by the authorities for GHG emissions during decision-making. In the United Kingdom the DECC set the shadow price of carbon at £60 in sectors not subject to the EU ETS. This value will rise progressively (reaching £200 in 2050). In France, the shadow price changes from 32 EUR per tonne of CO_2 in 2010 to 200 EUR in 2050 (a price range of 150 – 350 EUR).

Social cost of carbon: This is the theoretical carbon price at which the marginal cost of reduction is equal to the marginal benefit of reduction. The SCC enables the benefits of a given climate policy to be determined from the perspective of a cost–benefit analysis.

Stranded assets: assets that have suffered from unanticipated or premature write-downs, devaluations. Coal and other hydrocarbon resources may have the potential to become stranded because of stringent climate policies and the development of renewables.

Voluntary offsetting: Voluntary offsetting includes compensation initiatives for all or part of the emissions produced by an organization or an individual. Participants purchase and cancel carbon credits, provided there is no regulatory constraint.

United Nations Framework Convention on Climate Change (UNF-CCC): The UNFCCC (also known as the Climate Convention) was adopted at the Earth Summit in Rio and came into force in March 1994. Its aim is "the stabilization of greenhouse gas concentrations in the atmosphere at a level that would prevent dangerous anthropogenic interference with the climate system" (Article 2).

Bibliography

ADEME (2005). Introduction à l'analyse de cycle de vie (ACV), external briefing note.

ADEME (2010). Bilan carbone: guide des facteurs des émissions, version 6.

Arrhenius, S. (1896). On the influence of carbonic acid in the air upon the temperature of the ground. *Philosophical Magazine*, 41, 237-276.

Baumstark, L. (2007). La mesure de l'utilité sociale des investissements: l'enjeu du processus de production des valeurs tutélaires. In J. Maurice & Y. Crozet (Eds.), *Les dimensions critiques de calcul économique*. Paris: Economica.

Bellassen, V. and Leguet, B. (2007). The emergence of voluntary carbon offsetting, CDC Climat, research report No 11, www.cdcclimat.com/IMG/pdf/11_Etude_Climat_EN_Carbon_Neutrality.pdf.

Belzer, D.B. (2009). *Energy Efficiency Potential in Existing Commercial Buildings: Review of Selected Recent Studies*. Report, Washington, DC: US Department of Energy.

Biermann, F. and Boas, I. (2010). Preparing for a warmer world: towards a global governance system to protect climate refugees. *Global Environmental Politics*, 10(1), 60-88.

Blanchard, O. and Criqui, P. (2000). La valeur du carbone: un concept générique pour les politiques de réduction des émissions. *Économie internationale*, 82(2), 75-102.

Bodansky, D. (2010). The Copenhagen Climate Change Conference: a post-mortem. *American Journal of International Law*, 104, 230-240.

Bosquet, B. (2000). Environmental tax reform: does it work? A survey of the empirical evidence. *Ecological Economics*, 34(1), 19-32.

Braun, M. (2009). The evolution of emissions trading in the European Union: the role of policy networks, knowledge and policy entrepreneurs. *Accounting, Organizations and Society*, 34(3-4), 469-487.

Brohé, A. (2008). *Les Marchés de quotas de CO2*. Brussels: Larcier.

Brohé, A. (2013). *La comptabilité carbone*. Paris: La Découverte.

Brohé, A., Howarth, N. & Eyre, N. (2009). *Carbon Markets. An International Business Guide*. London: Earthscan.

Brohé, A. (2014). Whither the CDM? Investment outcomes and future prospects. *Environment, Development and Sustainability*, 16(2), 305-322.

Brouwer, R., Brander, L. & Van Beukering, P. (2008). "A convenient truth": air travel passengers' willingness to pay to offset their CO2 emissions. *Climatic Change*, 90(3), 299-313.

BSI (2008 and 2011). Publicly Available Specification (PAS) 2050. Specification for the Assessment of the Life Cycle Greenhouse Gas Emissions of Goods and Services.

BSI (2010). Publicly Available Specification (PAS) 2060. 2010 Specification for the Demonstration of Carbon Neutrality.

Callon, M. & Latour, B. (1981). Unscrewing the big Leviathan: how actors macro-structure reality and how sociologists help them to do so. In K.D. Knorr-Cetina & A.V. Cicourel (Eds.), *Advances in Social Theory and Methodology. Toward an Integration of Micro- and Macro-Sociologies* (pp. 277-303). Boston: Routledge/Kegan Paul.

Capron, M. & Quairel-Lanoizelée, F. (2010). *La Responsabilité sociale d'entreprise*, Paris: La Découverte.

Carbon Tracker (2013). Unburnable carbon 2013: wasted capital and stranded assets.

CDP (2015). Putting a price on risk: carbon pricing in the corporate world.

CGDD-SOeS (2010). CO_2 et activités économiques de la France: tendances 1990–2007 et facteurs d'évolution. *Études & Documents*, 27.

Chiquet, C. (2015). Variant 3: emissions of a company/institution rather than a site: the case of the Shenzhen ETS. In V. Bellassen & N. Stephan (Eds.), *Accounting for Carbon* (pp. 263-282). Cambridge: Cambridge University Press.

Commission Prada (2010). La régulation des marchés du CO_2. Rapport de la mission confiée à Michel Prada, Ministère de l'Économie, de l'Industrie et de l'Emploi, April 2010.

Cournede, B. & Gastaldo, S. (2002). Combinaison des instruments prix et quantités dans le cas de l'effet de serre. *Économie et Prévision*, 5(156), 51-62.

Crocker, T.D. (1966). The structuring of atmospheric pollution control systems. In H. Wolozin (Ed.), *The Economics of Air Pollution* (pp. 61-86). New York: W.W. Norton.

Dales, J.H. (1968). *Pollution Property and Prices*. Toronto: Toronto University Press.

DECC (Department of Energy and Climate Change) (2009, July). Carbon valuation in UK policy appraisal. A revised approach.

DECC (Department of Energy and Climate Change) (2015, 8 May). 2010 to 2015 government policy: Energy and climate change, evidence and analysis. Policy paper, https://www.gov.uk/government/publications/2010-to-2015-government-policy-energy-and-climate-change-evidence-and-analysis/2010-to-2015-government-policy-energy-and-climate-change-evidence-and-analysis

Delbosc, A. & Goubet, C. (2011). Panorama: carbon markets and prices around the world in 2011. In C. de Perthuis & P.A. Jouvet (Eds.), *Climate Economics in Progress 2011*. Paris: Economica.

EC (2000). COM (2000) 87 final. Green paper on greenhouse gas emissions trading within the European Union.

EIA (US Energy Information Administration) (2016). *International Energy Outlook 2016*. DOE/EIA-0484(2016) (May 2016), http://www.eia.gov/forecasts/ieo/pdf/0484(2016).pdf

EIA (US Energy Information Administration) (2016). *International Energy Outlook 2016*. Report. Washington, DC: US Energy Information Administration.

Ekins, P. & Speck, S. (Eds.). (2011). *Environmental Tax Reform (ETR). A Policy for Green Growth*.Oxford/New York: Oxford University Press.

Elbeze, J. & de Perthuis, C. (2011, 9 April). Vingt ans de taxation du carbone en Europe: les leçons de l'expérience. Les Cahiers de la chaire Économie du climat.

Engels, A. (2009). The European emissions trading scheme: an exploratory study of how companies learn to account for carbon. *Accounting, Organizations and Society*, 34(3–4), 488-498.

Etsy (2016). *The Etsy Sustainability Commission*, https://www.etsy.com/progress-report/2015/esc.

EU (2003a). Council Directive 2003/96/EC of 27 October 2003 restructuring the community framework for the taxation of energy products and electricity.

EU (2003b). Directive 2003/87/CE, of the European Parliament and of the Council of 13 October 2003 establishing a scheme for greenhouse gas emission allowance trading within the community and amending Council Directive 96/61/EC (1).

EU (2004a). 2004/156/EC, Commission Decision of 29 January 2004 establishing guidelines for the monitoring and reporting of greenhouse gas emissions.

EU (2004b). Directive 2004/101/CE, of the European Parliament and of the Council of 27 October 2004 establishing a scheme for greenhouse gas emission allowance trading within the community using the project mechanisms of the Kyoto Protocol.

EU (2007). Commission decision 2007/589/EC establishing guidelines for the monitoring and reporting of GHG emissions.

FEVE (2012). FEVE Full Life Cycle Inventory, www.feve.org.

Gerlagh, R. & Lise, W. (2005). Carbon taxes: a drop in the ocean, or a drop that erodes the stone? The effect of carbon taxes on technological change. *Ecological Economics*, 54, 241-260.

Gillenwater, M. (2012). What is additionality? Part 1: A long standing problem. Discussion Paper, Greenhouse Gas Management Institute, Silver Spring.

Glen, P., Minx, J., Weber, C. & Edenhofer, O. (2011). Growth in emission transfers via international trade from 1990 to 2008. Proceedings of the National Academy of Sciences.

Grubb, M. (1989). The greenhouse effect: negotiating targets. *International Affairs*, 66(1), 67-89.

Hamrick, K. & Goldstein, A. (2016). Raising ambition. State of the voluntary carbon markets 2016, Ecosystem Marketplace.

Haya, B. (2009). Measuring emissions against an alternative future: fundamental flaws in the structure of the Kyoto Protocol's clean development mechanism. Energy and Resources Group Working Paper, University of California, Berkeley.

Hope, C. (2005). Integrated assessment models. In D. Helm (Ed.), *Climate Change Policy* (pp. 77-98). Oxford: Oxford University Press.

IPCC (1999). *Aviation and the global atmosphere*. Cambridge: Cambridge University Press.

IPCC (2000a). *The IPCC good practice guidance and uncertainty management in national greenhouse gas inventories*.

IPCC (2000b). *Good practice guidance for land use, land-use change and forestry*. Report. Published by the Institute for Global Environmental Strategies (IGES) for the IPCC, Japan.

IPCC (2006). *2006 IPCC guidelines for national greenhouse gas inventories*.

IPCC (2007). *Climate change 2007*. Contribution of working groups I, II and III to the Fourth Assessment Report of the Intergovernmental Panel on Climate Change (core writing team: K.R. Pachauri and A. Reisinger (Eds.), Geneva: IPCC.

IPCC (2014). Climate change 2014. Contribution of working groups I, II and III to the Fifth Assessment Report of the Intergovernmental Panel on Climate Change (core writing team: K.R. Pachauri and L.A. Meyer (Eds.), Geneva: IPCC.

ISO (2006a). ISO 14040: 2006, Environmental management – Life cycle assessment – Principles and framework.

ISO (2006b). ISO 14044: 2006, Environmental management – Life cycle assessment – Requirements and guidelines.

ISO (2006c). ISO 14064-1: 2006, Greenhouse gases – Part 1: Specification with guidance at the organization level for quantification and reporting of greenhouse gas emissions and removals.

ISO (2006d). ISO 14064-2: 2006, Greenhouse gases – Part 2: Specification with guidance at the project level for quantification, monitoring and reporting of greenhouse gas emission reductions or removal enhancements.

ISO (2006e). ISO 14064-3: 2006, Greenhouse gases – Part 3: Specification with guidance for the validation and verification of greenhouse gas assertions.

ISO (2006f). ISO 14025: 2006, Environmental labels and declarations – Type III environmental declarations – Principles and procedures

ISO (2011). ISO/TS 14067-1: 2013 Carbon footprint of products – Requirements and guidelines for quantification and communication.

ISO (2013). ISO/TR 14069: 2013 Greenhouse gases (GHG) – Quantification and reporting of GHG emissions for organisations (Carbon footprint of organisations) – Guidelines for application of ISO 14064-1.

Jesus Saenz, M. (2012). Driving more efficient logistics networks through horizontal collaboration, www.scmr.com/article/driving_more_efficient_logistics_networks_through_horizontal_collaboration.

Joskow, P. & Schmalensee, R. (1998). The political economy of market-based environmental policy: the US acid rain program. *Journal of Law and Economics*, 41(1), 37-84.

Kander, A., Jiborn, M., Moran, D., & Wiedmann, T. (2015). National greenhouse-gas accounting for effective climate policy on international trade. *Nature Climate Change*, 5, 431-435.

Kebe, A., Bellassen, V. & Leseur, A. (2011). Voluntary carbon offsetting by local authorities: Practices and lessons. *Climate Report*, 29, CDC Climat.

Kesicki, F. & Strachan, N. (2011). Marginal abatement cost (MAC) curves: Confronting theory and practice. *Environmental Science & Policy*, 14(8), 1195-1204.

Kyoto Protocol to the United Nations Framework Convention on Climate Change (2007). Kyoto, 11 December 1997.

MacKenzie, D. (2009). Making things the same: gases, emission rights and the politics of carbon markets. *Accounting, Organizations and Society*, 34(3-4), 440-455.

Maréchal, K. (2007). The economics of climate change and the change of climate in economics. *Energy Policy*, 35(10), 5181-5194.

Markussen, P. & Svendsen, G.T. (2005). Industry lobbying and the political economy of GHG trade in the European Union. *Energy Policy*, 33(2), 245-255.

Mathers, J., Craft, E., Norsworthy, M. & Wolfe, C. (2014). *The Green Freight Handbook*. Report. New York: Environmental Defense Fund.

Meinshausen, M., Meinshausen, N., Hare, W., Raper, S.C.B., Frieler, K., Knutti, R., Frame, D.J. & Allen, M.R. (2009). Greenhouse-gas emission targets for limiting global warming to 2°C. *Nature*, 458(7242), 1158-1162.

Milanesi, J. (2010). Éthique et évaluation monétaire de l'environnement: la nature est-elle soluble dans l'utilité?. *VertigO*, 10(2).

Nakano, S., Okamura, A., Sakurai, N., Suzuki, M., Tojo J. and Yamano, N. (2009). "The measurement of CO_2 embodiments in international trade: evidence from the harmonised input-output and bilateral trade database", OECD Working Papers, no. DSTI/DOC(2009)3.

National Grid (2002). *Managing Energy Costs in Office Buildings*. Report, E Source Companies LLC.

Nielsen (2015). *The Sustainability Imperative*, www.nielsen.com/content/dam/nielsenglobal/dk/docs/global-sustainability-report-oct-2015.pdf.

Nordhaus, W. (1991). To slow or not to slow: the economics of the greenhouse effect. *Economic Journal*, 101(407), 920-937.

OECD (1972). Recommendation of the council on guiding principles concerning international economic aspects of environmental policies. document no. C(72)128, Paris.

OECD (2012). OECD environmental outlook to 2050. Paris: OECD.

Olsen, K.H. (2007). The clean development mechanism's contribution to sustainable development: a review of the literature. *Climatic Change*, 84(1), 59-73.

Paris agreement to the United Nations framework convention on climate (2015). Paris, 12 December 2015.

Pasquier, J.L. (2010, January). Les comptes physiques de l'environnement, une base pour de nouveaux indicateurs sur l'interface économie-environnement. Le cas des émissions de CO_2 de la France. *La Revue du Commissariat général au développement durable* (pp. 75-83).

Paterson, M. (2012). Who and what are carbon markets for? Politics and the development of climate policy. *Climate Policy*, 12(1), 82-97.

Peters-Stanley, M., Hamilton, K., Marcello, T. & Sjardin, M. (2011). Back to the future. State of the voluntary carbon markets 2011. *Ecosystem Marketplace & Bloomberg New Energy Finance*. Washington, DC/New York.

Pigou, A. (1920). *The Economics of Welfare*, New York: Macmillan.

Principles for Responsible Investment (2016). About the PRI, https://www.unpri. org/about.

Project Management Institute (2011). *The Bottom Line on Sustainability*. White Paper, Newtown Square, PA: Project Management Institute, Inc.

Quinet, A. (2008). La Valeur tutélaire du carbone, La Documentation française/ Centre d'analyse stratégique, Rapports et documents, Paris.

Rogelj, J., den Elzen, M., Höhne, N., Fransen, T., Fekete, H., Winkler, H., Schaeffer, R., Sha, F., Riahi, K., & Meinsaussen, M. (2016). Paris agreement climate proposals need a boost to keep warming well below 2°C. *Nature*, 534, 631-639.

Sartor, O. (2012). The EU ETS carbon price: To intervene, or not to intervene? *Climate Brief*, 12, CDC Climat.

Sausen, R., Isaksen, I., Grewe, V., Hauglustaine, D., Lee, D.S., Myhre, G., Kohler, M., Pitari, G., Schumann, U., Stordal, F., & Zerefos, C. (2005). Aviation radiative forcing in 2000: an update on IPCC (1999). *Meteorologische Zeitschrift*, 114, 555-561.

Shishlov, I., Bellassen, V. & Leguet, B. (2012). Joint implementation: a frontier mechanism within the borders of an emissions cap. *Climate Report*, 33, CDC Climat.

Skjærseth, J.B. and Wettestad, J. (2008). *EU Emissions Trading. Initiation, Decision-Making and Implementation*. Aldershot: Ashgate.

Speck, S. (2008). The design of carbon and broad-based energy taxes in European countries. In J.E. Milne (Ed.), *The Reality of Carbon Taxes in the 21st Century, Environmental Tax Policy Institute/Vermont Journal of Environmental Law* (pp. 31-60). South Royalton, VT: Vermont Law School.

Stern, N. (2006). *The Economics of Climate Change. Stern Review on the Economics of Climate Change*. Cambridge: Cambridge University Press.

Sutter, C. and Parreño, J.C. (2007). Does the current clean development mechanism (CDM) deliver its sustainable development claim? An analysis of officially registered CDM projects. *Climatic Change*, 84(1), 75-90.

UNCTAD (1992). Combating Global Warming. Study on a Global System of Tradable Carbon Emission Entitlements, United Nations Conference on Trade and Development, New York.

UNFCCC (1992). United Nations Framework Convention on Climate Change. New York, 9 May.

US Department of Agriculture Forest Service (2014, 29 April). Indicator 6.28: Total and per capita consumption of wood and wood products, www.fs.fed.us/research/sustain/criteria-indicators/indicators/indicator-628.php.

US Department of Energy (2016a). Driving more efficiently. 24 September. https://www.fueleconomy.gov/feg/driveHabits.jsp.

US EPA (Environmental Protection Agency) (2015, 8 October). Power Profiler, https://www.epa.gov/energy/power-profiler.

US SIF (2014). *US Sustainable, Responsible and Impact Investing Trends 2014*, www.ussif.org/trends.

Wara, M. (2007). Is the global carbon market working? *Nature*, 445, 595-596.

Weitzman, M.L. (1974). Prices vs. quantities. *Review of Economic Studies*, 41(4), 477-491.

Willard, B. (2012). *The New Sustainability Advantage: Seven Business Case Benefits of a Triple Bottom Line.* 10th ed., Gabriola Island, British Columbia: New Society Publishers.

World Bank (2012). States and Trends of the Carbon Market. May. Washington, DC.

World Bank (2014). Electric power transmission and distribution losses (% of output), http://data.worldbank.org/indicator/EG.ELC.LOSS.ZS.

WRI and WBCSD (2001). The Greenhouse Gas Protocol. A Corporate Accounting and Reporting Standard, 2004, revised version.

WRI and WBCSD (2011a). The Greenhouse Gas Protocol. Corporate Value Chain (Scope 3) Accounting and Reporting Standard.

WRI and WBCSD (2011b). Product Life-Cycle Accounting and Reporting Standard.

WRI and WBCSD (2013). The Greenhouse Gas Protocol. A Corporate Accounting and Reporting Standard, 2013, revised version.

WRI and WBCSD (2014). Mitigation Goal Standard.

About the author

Arnaud Brohé, PhD, is CEO of CO_2logic Inc., an international carbon reduction and offset firm. He is a renowned expert in the field of climate and energy policies (carbon markets, carbon audits, support schemes for renewable energy and energy efficiency), environmental management (energy, water, waste and biodiversity impacts) and sustainable reporting. Brohé teaches  at *Université libre de Bruxelles* and Imperial College London, and his book *Carbon Markets*, published by Routledge in 2009, was the winner of the Choice Outstanding Academic Award (2010).

For Product Safety Concerns and Information please contact our EU
representative GPSR@taylorandfrancis.com Taylor & Francis Verlag GmbH,
Kaufingerstraße 24, 80331 München, Germany

Printed and bound by CPI Group (UK) Ltd, Croydon, CR0 4YY
12/05/2025
01867601-0001